Subjective Probability Models for Lifetimes

Fabio Spizzichino

Department of Mathematics
University La Sapienza
Rome Italy

CRC Press
Taylor & Francis Group
Boca Raton London New York

CRC Press is an imprint of the
Taylor & Francis Group, an **informa** business

A CHAPMAN & HALL BOOK

To Daniela, Valeria and Flavia

Contents

Contents

Preface

From a mathematical point of view, the subject of this monograph can be described as a study of exchangeable, non-negative random variables or, in other words, of symmetric probability measures over $\mathbb{R}_+^n$.

However, the interest is essentially in the role of conditioning in lifetime models and, more than on analytical aspects, the focus is on some ideas related to applications, especially in reliability and survival analysis; in fact the random variables of interest, $T_1, ..., T_n$, have the meaning of lifetimes of different units (or individuals) and most of our attention will be focused on conditional "survival" probabilities of the type

$$P\{T_{h+1} > s_{h+1}, ..., T_n > s_n | T_1 = t_1, ..., T_h = t_h, T_{h+1} > r_{h+1}, ..., T_n > r_n\}.$$

In particular we study notions of *dependence* and notions of *aging*, which provide the tools to obtain inequalities for those conditional probabilities.

The monograph is addressed to statisticians, probabilists, and engineers, interested in the methods of Bayesian statistics and in Bayesian decision problems; the aim is to provide a conceptual background which can be useful for sound applications of such methods, mostly (but not exclusively) in reliability and survival analysis.

Since the Bayesian approach is mainly motivated by a subjectivist interpretation of probability, our study will be developed within the frame of *subjective probability*, where the probability of an event is interpreted as a degree of belief, related with a personal state of information; within such a frame, probability is indeed a tool for describing a state of information (or a state of uncertainty). A characterizing feature of subjective probability is, in fact, that any unknown quantity is treated as a random variable (i.e., there is no room for "deterministic but unknown quantities") and uncertainty about it is expressed by means of a probability distribution.

This entails that a probability assessment explicitly depends on the information level. In particular, as a basic point of our discussion, we have to point out that aging and dependence properties for lifetimes depend on the actual flow of available information.

In general an approach based on such general lines turns out to be quite flexible and convenient; as a matter of fact it provides a natural setting for those applications in which different information levels must be considered and compared, as can be the case in the problems considered here.

On the other hand, the same approach involves, for many probabilistic notions, an interpretation completely different from that in other approaches and, as we shall discuss, this has some effects even on the meaning of the notions of dependence and of aging. In particular notions of stochastic dependence are to be understood in a special way, since dependence can originate both from "physical" interactions and from situations in which learning is present.

Since ignoring such effects can be a source of logic difficulties in the application of Bayesian methods, we shall extensively discuss aspects of dependence, of aging, and of their mutual relations, in the frame of subjective probability.

As I mentioned above, the discussion will be limited to the case of exchangeable lifetimes. For such a case, special notions such as multivariate hazard rates functions, notions of aging and dependence, which arise in reliability and survival analysis, will be discussed in detail.

The choice of limiting the study to exchangeable lifetimes has the following motivations:

- it allows us to analyze conceptual points related to the role of *information*, picking off those aspects which are not essential in the present discussion;

- exchangeability is, in my opinion, the natural background needed to clarify relations between concepts of aging and dependence in the subjective standpoint;

- exchangeability provides an introduction and a general frame for a topic which will be discussed in detail: distributions with *Schur-concavity* or *Schur-convexity* properties and related role in the analysis of failure data (such properties are defined in terms of the concept of *majorization* ordering).

The latter type of distributions describe in fact a very particular case of exchangeability. However, they are of importance both from a conceptual and an application-oriented point of view, in that they give rise, in a sense, to a natural generalization of the fundamental case of independent or conditionally independent (identically distributed) lifetimes with monotone failure rate.

Special cases of distributions with Schur properties are the distributions with *Schur-constant* joint survival functions or *Schur-constant* densities. These constitute natural generalizations of the basic cases of independent or conditionally independent (identically distributed) exponential lifetimes. Like the exponential distribution is the basic and idealized probability model for standard reliability methods, similarly Schur-constant densities can be seen as the idealized models in the setting of multivariate Bayesian analysis of lifetimes. In fact, densities with Schur properties and exchangeable densities could be, respectively, seen, in their turn, just as the most natural generalizations of Schur-constant densities; furthermore a suitable property of *indifference with respect to aging*, or of *no-aging*, owned by the latter, is an appropriate translation of the *memory-less* property of exponential distributions, into the setting considered here. We shall see (in Chapter 4) that no-aging is substantially a property of exchangeability for residual lifetimes of units of different ages.

The monograph consists of five chapters.

The first chapter provides a background to the study of subjective probability and Bayesian statistics. In particular we shall discuss the impact of subjective probability on the language and formalization of statistics and the related role of exchangeable random variables.

In the second chapter, by concentrating attention on the case of exchangeability, we analyze fundamental notions of multivariate probability calculus for non-negative random variables, such as *survival functions, conditional survival probability given histories of failures and survivals, one-dimensional hazard rate functions,* and *conditional multivariate hazard rate functions.*

Based on the fundamental concepts of stochastic comparisons, the third chapter is devoted to the presentation of some notions of stochastic dependence, aging and to their mutual relations, with an emphasis on the exchangeable case.

The fourth chapter will focus on the probabilistic meaning of distributions with Schur-constant, Schur-concave and Schur-convex survival functions or density functions. After discussing these notions, we shall illustrate related properties and some applications to problems in survival analysis.

Chapter 5 will be devoted to Bayes decision problems; two main features of this chapter are the following:

i) some typical problems in the field of reliability applications are given the shape of Bayes decision problems, special attention being paid to life-testing and burn-in problems;

ii) it is shown how different concepts of stochastic orderings, dependence and majorization enter in the problem of obtaining inequalities for Bayes decisions when different sets of observed data are compared.

In order to limit the mathematical difficulties, all these arguments will be essentially treated under the assumption that the joint distributions of lifetimes admit a joint density function.

My intention was to give each separate chapter a specific character and identity of its own and then an effort was made to maintain a reasonable amount of independence among chapters; however, the same group of a few basic models is reconsidered several times between Chapters 2 and 5, to provide examples and comparisons at different stages of the treatment.

For the sake of independence among chapters, selective lists of bibliographical references will be presented at the end of each of them. Such lists are far from being exhaustive, and I apologize in advance for the omissions which they surely present.

On the other hand, these lists may be still too wide and the reader may desire to have an indication of a more basic literature. For this reason a reduced list of really basic, by-now classic, books, is presented in the essential bibliography at the end of the monograph. Only a very small part of all ideas and mathematical material contained in those books was sufficient to provide the basic background for the arguments developed here and, of course, the reader is not at all assumed to be familiar with all of them.

It is assumed, rather, a background on calculus in several variables, basic theory of probability at an intermediate level, fundamentals of Bayesian statistics, basic elements of stochastic processes, reliability and life testing at an introductory level.

As far as Bayesian statistics is concerned, only the knowledge of the basic language and of Bayes' formula is assumed (the notation of Stiltjes integrals will be sometimes used). As to the reliability background, the reader should be familiar with (and actually interested in the use of) the most common concepts of univariate aging, such as IFR, DFR, NBU, etc.

No previous knowledge of the two basic topics of exchangeability and stochastic orderings is strictly assumed.

The project of this monograph evolved over quite a long period, somehow changing in its form, over years; such an evolution was assisted, at its different stages, by the project managers at Chapman & Hall, previously, and at CRC Press, lately; I would like to thank them for assistance, encouragement, and for their professional commitment.

Special thanks are also due to Marco Scarsini, for encouragement and suggestions that he has been providing since the early stage of the project.

The basic kernel of ideas presented here developed over years, originating from several, long discussions that I had with Richard E. Barlow and Carlo A. Clarotti, about ten years ago. Essentially our interest was focused on the meaning of positive and negative aging and on the analysis of conditions which justify procedures of burn-in, for situations of stochastic dependence. On related topics, I also had a few illuminating discussions with Elja Arjas and Moshe Shaked.

Several ideas presented here also developed as a consequence of discussions with co-authors of some joint papers and with students of mine, preparing their *Tesi di Laurea* in Mathematics, at the University "La Sapienza" in Rome.

At an early stage of the preparation of the manuscript, helpful comments were provided by Menachem Berg, Julia Mortera, Giovanna Nappo and Richard A. Vitale. In the preparation of the most recent version, a big help and very useful comments came from several colleagues, among whom, in particular, are Richard E. Barlow, Bruno Bassan, Uwe Jensen, Giovanna Nappo, Ludovico Piccinato, Wolfgang Runggaldier, Marco Scarsini, and Florentina Petre.

I am grateful to Bruno and Florentina also for their help on matters of technical type: in particular Bruno introduced me to the pleasures (and initial sorrows) of Scientific Word and Florentina helped me with insertion of figures. Several grants, from C.N.R. (Italian National Council for Research) and M.U.R.S.T. (Italian Ministry for University and Scientific Research), supported my research activity on the topics illustrated in the monograph, and are here gratefully acknowledged.

Rome, March 2001.

Notation and Acronyms

Notation

$E_1, E_2, \ldots$ random events

$\mathbf{1}_E$ indicator of the event E

$\mathcal{E}$ family of exchangeable events

$X_1, X_2, \ldots$ random variables

$\mathcal{L}(X)$ probability law of X

$S_n \equiv \sum_{i=1}^{n} X_i$, for exchangeable random variables $X_1, X_2, \ldots$

$\omega_k^{(n)} \equiv P\{S_n = k\}$, with $X_1, X_2, \ldots$ binary exchangeable random variables

$\omega_k \equiv \omega_k^{(k)} = P\{S_k = k\}$

$p_k^{(n)} \equiv \dfrac{\omega_k^{(n)}}{\binom{n}{k}}$

N_0 maximum rank of a family of exchangeable events

$\mathbb{R}$ set of the real numbers

$\mathbb{R}_+$ set of the non-negative real numbers

$[x]_+ \equiv \begin{cases} x & \text{if } x \geq 0 \\ 0 & \text{if } x < 0 \end{cases}$

$\mathbb{E}(X)$ expected value of the random variable X

$X_{(1)}, \ldots, X_{(n)}$ order statistics of random variables $X_1, X_2, \ldots, X_n$

$F^{(n)}(x_1, \ldots, x_n) \equiv P\{X_1 \leq x_1, X_2 \leq x_2, \ldots, X_n \leq x_n\}$, joint distribution function of n exchangeable random variables $X_1, X_2, \ldots, X_n$

$f^{(n)}(x_1, \ldots, x_n)$ joint density function of n exchangeable random variables

$f^{(n)}(t_1, \ldots, t_n)$ joint density function of n non-negative exchangeable random variables (lifetimes)

$U_1, \ldots, U_n$ units or individuals

$T_1, T_2, ..., T_n$ exchangeable lifetimes

$s_1, ..., s_n$ set of *ages*

$\overline{F}^{(n)}(s_1, ..., s_n) \equiv P\{T_1 > s_1, T_2 > t_2, ..., T_n > s_n\}$, joint survival function of n exchangeable lifetimes

$r(t)$ (one-dimensional) failure (or hazard) rate function

$R(s) \equiv \int_0^s r(t)\, dt$ cumulated hazard rate function

D observed data (possibly containing *survival* data)

D' *dynamic history* of the form $\{T_{i_1} = t_1, ..., T_{i_h} = t_h, T_{j_1} > t, ..., T_{j_{n-h}} > t\}$

$\mathfrak{h}_t$ *dynamic history* of the form $\{T_{(1)} = t_1, ..., T_{(h)} = t_h, T_{(h+1)} > t\}$ or

$\{\mathbf{X}_I = \mathbf{x}_I; \mathbf{X}_{\bar{I}} > t\mathbf{e}\}$

Θ *parameter* in a parametric statistical model

W *state of nature* in a Bayes decision problem

Π_0 initial (or a-priori) distribution for a parameter Θ or for a state of nature W

π_0 density of Π_0 (when it does exist)

$\Pi(\cdot|D), \pi(\cdot|D)$ conditional distribution and conditional density for Θ or W, given the observed data D

$\lambda_t^{(n)}, \lambda_t^{(n-h)}(t_1, ..., t_h)$ *multivariate conditional hazard rate* functions for exchangeable lifetimes $T_1, ..., T_n$

$\lambda_{i|I}(t|\mathbf{x}_I)$ *multivariate conditional hazard rate* functions for non-necessarily exchangeable lifetimes

$H_t \equiv \sum_{j=1}^n \mathbf{1}_{[T_j \leq t]}$, stochastic process counting the number of failures observed up to time t for n units with exchangeable lifetimes $T_1, ..., T_n$

$a \wedge b \equiv \min(a, b); \; a \vee b \equiv \max(a, b)$

$Y_t \equiv \sum_{i=1}^n (T_{(i)} \wedge t)$ *Total Time on Test* (TTT) process for n units with exchangeable lifetimes $T_1, ..., T_n$

$Z_t \equiv (H_t, Y_t)$

$Y_{(h)} \equiv \sum_{i=1}^{h} T_{(i)} + (n-h) \cdot T_{(h)}$ *Total Time on Test cumulated* up to the h-th failure, $h = 1, 2, ..., n$

$C_h \equiv (n-h+1) \cdot \left(T_{(h)} - T_{(h-1)}\right)$, $h = 1, ..., n$ *normalized spacings between order statistics*

$\mathcal{H}_t \equiv (H_t, T_{(1)} \wedge t, ..., T_{(n)} \wedge t)$ *history process*

$\leq_{st}, \leq_{hr}, \leq_{lr}, \leq_{ch}$ one-dimensional or multidimensional stochastic orderings

$\prec, \succ$ majorization orderings

A action space

l loss function

δ decision function (or strategy)

$a_{\mathcal{S}}^{(n-h)}(t_1, t_2, ..., t_h)$ residual duration of a burn-in procedure, after h failures, according to the burn-in strategy $\mathcal{S}$.

Acronyms
a.s. almost surely
i.i.d Independent and identically distributed (random variables)
m.c.h.r. Multivariate conditional hazard rate (functions)
TTT Total time on test
st stochastic (ordering)
hr hazard rate (ordering)
lr likelihood ratio (ordering)
TP Totally positive
MTP Multivariate totally positive
PC Positively correlated
PUOD Positively upper orthant dependent
HIF Hazard increasing upon failure
SL Supporting lifetimes
WBF Weakened by failures
IFR Increasing failure rate
DFR Decreasing failure rate
NBU New better than used
NWU New worse than used
IFRA Increasing failure rate in average
PF Polya frequency

MRR Multivariate reverse regular
NUOD Negatively upper orthant dependent
MIFR Multivariate increasing failure rate
MPF Multivariate Polya frequency
OLFO Open-loop feedback optimal.

Chapter 1

Exchangeability and subjective probability

1.1 Introduction

In this introductory section we want to summarize briefly the main, well-known differences between the subjectivist and frequentist interpretations of probability. Then we sketch the implications that, in the formalism of statistics, are related to that.

We aim mainly to stress aspects that sometimes escape the attention of applied Statisticians and Engineers interested in Bayesian methods.

For this purpose we limit ourselves to a concise and informal treatment, leaving a more complete presentation to the specialized literature.

In the subjective approach, *randomness* is nothing but a lack of information. A random quantity is any unambiguously defined quantity X which, according to your state of information, takes on values in a specified space $\mathcal{X}$ but your information is such that you are not able to claim with certainty which element $x \in \mathcal{X}$ is the "true" value of X.

A random event E is nothing but a statement about which you are not sure: you are not able to claim with certainty if E is true or false.

Personal probability of E is your *degree of belief* that E is true.

We start from the postulate that you can assess a degree of belief for any event E of interest in a specified problem: your state of partial information about E is described by the assessment of a degree of belief. Similarly, the state of partial information about a random quantity X is described by a probability distribution for X. Generally a state of partial information is described by a probability distribution on a suitable family of events and, in turn, a probability distribution is induced by a state of information, so that the (personal) probability distribution and the state of partial information are two different

1

aspects of the same object.

Roughly speaking, in the frequentist approach probability has the meaning of *frequency of successes* in a very large number of analogous trials; i.e. we can define probability only with reference to a collective of events and it is senseless to speak of probability for a singular event (this is an event which cannot be reasonably embedded in some collective).

More generally, in the frequentist approach, we must distinguish between random variables (which can be embedded in a collective of observations) and deterministic quantities with an unknown value (these are originated from singular experiments); for the latter it is impossible to define a (frequentist) probability distribution.

Let us now underline some fundamental aspects implicit in the subjective approach; these make the latter approach substantially different from the frequentist one and also make clear that the language of statistics is to be radically modified when switching from one approach to the other.

1) Personal probability of an event is by no means a quantity with a universal physical meaning, intrinsically related to the nature of the event; it changes when the state of information changes and, obviously, depends on the individual who assesses it.

2) It is senseless to speak of a probability which is unknown to the individual who is to assess it (even though he/she may have some difficulty in stating it precisely).

This is a focal point and we aim to explain it by means of the following argument.

Think of the situations when the assessment of the conditional probability $P(E|x)$, of a specified event E given the knowledge of the value x taken by some quantity X, is, let us say, straightforward.

When the choice of X is appropriate, this can often happen, due, for example, to reasons of symmetry, or to large past experience, or to reasons of "intersubjective agreement" and so on.

For such situations, consider now the assessment of the probability of E, when the value of X is unknown to you.

The basic point in the subjective frame is that it is senseless to claim that $P(E|x)$ is the "true probability" but it is "unknown" due to ignorance about X.

In such cases X, being an unknown quantity, is to be looked at as a random quantity and a probability distribution $F_X(x)$ is to be assessed for it.

In order to derive the personal probability of E, you must rather "uncondition" with respect to X.

This means that the probability of the event E is to be computed according to the rule

$$P(E) = \int P(E|x)dF_X(x).$$

3) Stochastic interdependence between different events (or different random quantities, in general) can exist even if they are deemed to be physically independent.

This usually happens when we "uncondition" with respect to some unknown quantity, in the sense specified in the point above.

In such cases the knowledge of the values taken by some of the considered events modifies the state of information concerning those events which have not yet been observed.

This is the case of interdependence due to an *increase of information* and it is of fundamental importance in the problems of statistics.

This topic will have in particular a crucial role in what follows and we shall come back to it several times.

A special case of that is met when dealing with the form of dependence created by situations of conditional independence; such situations will be often considered in this monograph and a basic case of interest is the one described in Example 2.1.

4) All events and all unknown quantities are singular in the subjective approach; in some cases there is a situation of similarity (or *symmetry*), relative to the state of information of a given individual; this is formalized by means of the concept of exchangeability.

5) Can personal probability, not being an observable physical quantity and depending on an individual's opinion, be of some use in real practical problems?

The answer is "Yes": it is to be employed in decision problems under uncertainty.

By combining personal probabilities with utilities, one can check if a decision procedure is a *coherent* one; only coherent procedures are to be taken into consideration.

Coherency is a fundamental concept in the subjective approach and the (Kolmogorov-like) axioms of probability are interpreted and justified in terms of coherency.

6) As mentioned, the state of information about a random quantity X is declared by means of a probability distribution.

The state of information about X, after observing the value y taken by another quantity Y will in general be a modified one and then declared by using a different probability distribution: by the constraint of coherency, the latter must coincide with the *conditional distribution of X given* $(Y = y)$.

In "regular" cases, the latter distribution can be obtained from the former by use of Bayes formula.

7) By the very nature of personal probability, it makes sense to assess probability distributions only for those quantities which have a clear physical meaning and are of real interest in the problem at hand.

Let us now mention some immediate consequences that the above points imply in the treatment of statistics.

a) In orthodox statistics, *parameters* normally are looked on as deterministic quantities with an unknown "true" value; such a value affects the (frequentist) probability distributions of statistical observations.

The latter are random variables. Even though the statistician may have some *a-priori* information about the value of the parameter, no systematic method is available to formalize how to deal with it.

In Bayesian statistics, any unknown quantity is treated as a random variable and then there is a sort of symmetry between parameters and statistical observations.

The parameter, being a random quantity of its own, follows a (personal) probability distribution; the latter depends on the statistician's state of information: thus we have a prior distribution on the parameter (describing the state of information before taking observations) and a posterior distribution (describing the state of information after taking observations).

The terms *prior* and *posterior* only have a conventional meaning (for instance a prior can be obtained as a posterior following some previous observations). However it is to be said that, in view of what was mentioned at point 7) above, one often tries to avoid introducing a parameter, as we shall illustrate later on.

b) The easiest and most fundamental case of interest in statistics is defined, in the orthodox statistics language, by statistical observations which form a sequence of Bernoulli trials (i.e. independent, equiprobable events) with an "unknown" (frequentist) probability; the latter is the parameter which we must estimate on the basis of the observed events.

In order to characterize and to study this situation in a subjectivist standpoint, we need to introduce a suitable concept; indeed the observable events are not stochastically independent (since the observation of some carries a piece of information about others); furthermore subjective probability cannot coincide with the (unknown) frequentist probability.

Last, but not least, we must explain, in subjectivist terms, what the frequentist probability (parameter to be estimated) is.

The above leads us to introduce the concept of (a family of) exchangeable events. A natural generalization is the concept of (a family of) exchangeable random quantities which is introduced to render the concept of random sample in the subjectivist language.

c) Sometimes the parameter in a statistical model has a clear physical meaning of its own.

In those cases the very aim of statistical analysis is to study how statistical observations modify the state of information about the parameter itself.

In such cases there is a certain analogy between the orthodox and the Bayesian approach: the parameter is to be estimated based on the observed values taken by statistical variables; the differences between the two approaches essentially lie in the methods used for the estimation of parameters, once the values taken by statistical variables have been observed.

In the Bayesian approach, the posterior distribution which is obtained by conditioning with respect to the observations constitutes the basis for the estimation procedures.

A very important consequence is that, at least in regular cases, the *likelihood principle* must hold: since the posterior distribution only depends on the prior distribution and on the likelihood function associated with the observed results, it follows that two different results giving rise to the same likelihood (up to some proportionality constant, possibly) also give rise to the same posterior distribution for the parameter and then to the same estimation or decision procedure.

d) More often, however, the parameter in a (orthodox) statistical model merely has a conventional meaning, simply having the role of an index for the distributions of statistical variables.

In the subjectivist approach to Bayesian statistics this kind of situation is seen as substantially different from the one described formerly.

In this situation indeed it is recognized that the actual aim of the statistical analysis is a *predictive* one: on the basis of already observed statistical variables we need to predict the behavior of not yet observed variables rather than to estimate the "true" value of an hypothetical parameter.

Prediction can be made once the joint (personal) probability of all the conceivable observations (past and future) is assigned.

In these cases then attention is focused on the probability distribution obtained after "unconditioning" with respect to the unknown parameters.

Such distributions are usually called *predictive distributions.*

Problems of statistical inference and decisions formulated in terms of such distributions are said to be *predictive inference* problems and *predictive decision* problems, respectively.

Certain natural questions may then be put:

What is the relationship between the joint (personal) distribution of observables and the (frequentist) model?

Can the parameter be given a convincing Bayesian interpretation?

These problems are, in some cases, solved through de *Finetti-type theorems.* This will be briefly explained in Section 1.4.

In the next sections of this chapter we shall dwell in some detail on exchangeable events and exchangeable random variables. This will serve as a basis for what is to be developed in the following. Furthermore this will allow us to clarify better the points quoted above.

In the special framework of exchangeability, in particular, we shall briefly discuss the spirit of de Finetti-type results.

These results, which are useful to explain what a "parameter" is, can be seen from a number of different points of view. As we shall mention, a possible point of view is based on the role of (predictive) sufficient statistics.

Although such topics have a theoretical character, they will be important for a better comprehension of basic aspects related to the analysis of lifetimes.

If not otherwise specified, the term *probability* will stand for *subjective probability* of an individual (perhaps you).

1.2 Families of exchangeable events

We are interested in the probabilistic description of situations where, according to our actual state of information, there is indifference among different events.

Example 1.1. In a factory, a lot of n units of the same type has just been received and not yet inspected. We expect that there are a few defective units in the lot. Before inspection, however, we have no reason to distinguish among different units, as far as their individual plausibility to be defective is concerned: given $0 \leq k \leq m < n$, whatever group of m out of the n units has the same probability to contain k defective and $(m - k)$ good units.

The notion of exchangeability, for a family of events, formalizes such an idea of indifference and allows us to understand which are the possible cases of indifference among events.

Situations of indifference among (non-binary) random quantities will be studied in the next section, by means of the more general concept of exchangeable random variables.

Let $\mathcal{E} \equiv \{E_1, ..., E_n\}$ be a finite family of random events; the symbol X_i will denote the indicator of E_i, i.e. X_i is the binary random quantity defined by

$$X_i = \begin{cases} 1 & \text{if } E_i \text{ is true} \\ 0 & \text{if } E_i \text{ is false} \end{cases}.$$

Definition 1.2. $\mathcal{E}$ is a family of *exchangeable events* if, for any permutation $j_1, j_2, ..., j_n$ of $1, 2, ..., n$, $(X_1, ..., X_n)$ and $(X_{j_1}, ..., X_{j_n})$ have the same joint distribution: for any $1 \leq h < n$

$$P(X_1 = 1, ..., X_h = 1, X_{h+1} = 0, ..., X_n = 0) =$$

$$P(X_{j_1} = 1, ..., X_{j_h} = 1, X_{j_{h+1}} = 0, ..., X_{j_n} = 0). \tag{1.1}$$

For an arbitrary family of events $\mathcal{E} \equiv \{E_1, ..., E_n\}$, fix $1 \leq m \leq n$ and consider the probability of all successes over a fixed m-tuple of events $E_{j_1}, ..., E_{j_m}$:

$$p_{j_1, j_2, ..., j_m} \equiv P(E_{j_1} \cap ... \cap E_{j_m}).$$

Obviously, $p_{j_1, j_2, ..., j_m}$ will in general depend on the choice of the indexes $j_1, j_2, ..., j_m$; but, in the case of exchangeability, one can easily prove the following

Proposition 1.3. *Let $\mathcal{E} \equiv \{E_1, ..., E_n\}$ be a family of exchangeable events. Then, for any $1 \leq m \leq n$, $p_{j_1, j_2, ..., j_m}$ is a quantity depending on m but independent of the particular choice of the m-tuple $j_1, j_2, ..., j_m$: we can find n numbers $0 \leq \omega_n \leq \omega_{n-1} \leq ... \leq \omega_2 \leq \omega_1 \leq 1$ such that*

$$p_{j_1, j_2, ..., j_m} = \omega_m.$$

From now on in this section, $\mathcal{E} \equiv \{E_1, ..., E_n\}$ is a family of exchangeable events.

For $m = 1, 2, ..., n$, $1 \leq h \leq m - 1$, we shall use the notation

$$p_0^{(m)} \equiv P(X_1 = 0, ..., X_m = 0), \quad p_m^{(m)} \equiv P(X_1 = 1, ..., X_m = 1) = \omega_m.$$

and, for $1 \leq h \leq m - 1$,

$$p_h^{(m)} \equiv P(X_1 = 1, ..., X_h = 1, X_{h+1} = 0, ..., X_m = 0).$$

It will be also convenient to use the symbol $\omega_h^{(m)}$ to denote the probability of exactly h successes among any m events from $\mathcal{E}$: we let, for $1 \leq h < m < n$,

$$\omega_h^{(m)} \equiv P\{\sum_{i=1}^{m} X_i = h\} = \binom{m}{h} p_h^{(m)}. \tag{1.2}$$

As is immediate to check, it must be

$$p_h^{(m)} = p_h^{(m+1)} + p_{h+1}^{(m+1)}. \tag{1.3}$$

It is then self-evident that, for $1 \leq h \leq m < n$, $\omega_h^{(m)}$ can be computed in terms of $\{\omega_k^{(n)}, \ k = 0, 1, ..., n\}$. Later on we shall see the explicit form of the relation between $\{\omega_k^{(n)}\}_{k=1,...,n}$ and $\{\omega_n^{(m)}\}_{k=1,...,m}$, $m < n$ (Proposition 1.9 and Remark 1.12).

In what follows we shall present some relevant examples of exchangeable families, which will clarify fundamental aspects of the definition above.

Example 1.4. (Bernoulli scheme; binomial probabilities). Let $E_1, ..., E_n$ be judged to be mutually independent and such that $P(E_i) = p$ ($0 < p < 1$). It is immediate to check that $E_1, ..., E_n$ are exchangeable and

$$\omega_h = p^h, \quad \omega_h^{(m)} = \binom{m}{h} p^h (1 - p)^{m-h} \quad \text{for } 1 \leq h < m \leq n$$

To introduce the related next example, it is convenient to think in particular of the scheme of random drawings from an urn. An urn contains $M_1 = \theta M$ green balls and $M_2 = (1 - \theta)M$ red balls, where θ $(0 < \theta < 1)$ is a quantity known to you.

We perform n consecutive random drawings with replacement and set

$$E_i \equiv \{\text{a green ball at the i-th drawing}\}, (1 \leq i \leq n).$$

This obviously is a special case of the Bernoulli scheme $(p = \theta)$. In drawings with replacement it is immaterial whether M is known or unknown.

Example 1.5. (Random drawings with replacement from an urn with unknown composition; mixture of binomial probabilities).

Let us consider the urn as in the example above; but let us think of the case when θ is unknown to you.

θ is the value taken by a random quantity Θ and your prior state of information on Θ is described by a probability distribution Π_0 on the interval $[0, 1]$. Your personal probability is

$$P(E_i) = \int_0^1 \theta d\Pi_0(\theta).$$

Moreover, for $i \neq j$,

$$P(E_i \cap E_j) = \int_0^1 \theta^2 d\Pi_0(\theta) \, ,$$

whence $E_1, ..., E_n$ are not (subjectively) stochastically independent, in that

$$P(E_i \cap E_j) \neq P(E_i)P(E_j).$$

However, they are exchangeable; in particular they are *conditionally independent* given Θ and it is

$$\omega_h^{(m)} = \int_0^1 \binom{m}{h} \theta^h (1 - \theta)^{m-h} d\Pi_0(\theta).$$

Example 1.6. (Several tosses of a same coin). We toss the same coin n times and let $E_i \equiv \{\text{head at the i-th toss}\}$. It is natural to assume

$$P(E_1) = P(E_2) = ... = P(E_n).$$

Two different cases are now possible: you assess stochastic independence among $E_1, ..., E_n$ (this is a very strong position of yours) or you admit interdependence (this may be more reasonable), i.e. you admit there are results in the first $(h-1)$ tosses which may lead you to assess a conditional probability, about the outcome of the h-th toss, which differs from the initial one. Only in the first case we are in a Bernoulli scheme $(\omega_k = (\omega_1)^k)$; in any case, however, it is natural to assume that $E_1, ..., E_n$ are exchangeable.

Example 1.7. (Random drawings without replacement from an urn with known composition; hypergeometric probabilities). Let us go back to the example of the urn, containing $M_1 = \theta M$ green balls and $M_2 = (1 - \theta)M$ red balls, where both θ and M are known quantities. Consider the events $E_1, ..., E_n$, where

$$E_i \equiv \{\text{green ball at the i-th drawing}\}.$$

This time we perform n random drawings without replacement. $E_1, ..., E_n$ are such that $P(E_1) = ... = P(E_n) = \theta$.

They are not independent; however they are exchangeable and one has

$$\omega_k = \frac{M_1}{M} \frac{M_1 - 1}{M - 1} \frac{M_1 - 2}{M - 2} \cdots \frac{M_1 - k + 1}{M - k + 1},$$

$$\omega_h^{(m)} = \frac{\binom{M_1}{h}\binom{M_2}{m-h}}{\binom{M}{m}}, \quad max(0, m - M_2) \le h \le min(m, M_1). \tag{1.4}$$

Example 1.8. (Random drawings without replacement from an urn with unknown composition; mixture of hypergeometric probabilities). Consider the case of the previous example with θ unknown; then M_1 and M_2 are unknown too. Also in this case $E_1, ..., E_n$ are exchangeable and, by (1.4) and the rule of total probabilities, one obtains

$$\omega_h^{(m)} = \sum_{k=h}^{M-m+h} \frac{\binom{k}{h}\binom{M-k}{m-h}}{\binom{M}{m}} P\{M_1 = k\}. \tag{1.5}$$

Of course we shall understand $\binom{b}{a} = 0$, if $a > b$.

We shall see soon (Remark 1.12) that the situation contemplated in Example 1.8 can be used as a representation of the most general case of finite families of exchangeable events.

Before we analyze further, some general facts.

For an arbitrary exchangeable family $\{E_1, ..., E_n\}$, put

$$S_m = \sum_{i=1}^{m} X_i \; (m = 1, 2...., n)$$

so that Equation (1.2) becomes

$$P\{S_n = k\} = \omega_k^{(n)} = \binom{n}{k} p_k^{(n)}, \; k = 0, 1, ..., n.$$

Some relevant properties of the sequence $\{S_1, ..., S_n\}$ directly follow from the very definition of exchangeability.

Due to these properties, $\{S_1, ..., S_n\}$ has a fundamental role in the present study and it is the prototype for a general concept which will be introduced in Section 1.4. Fix $h \leq m < n$; first, it is

$$P\{S_m = h|S_n = k\} =$$

$$\frac{P\{(S_m = h) \cap (\sum_{i=m+1}^{n} X_i = k - h)\}}{P\{S_n = k\}} =$$

$$\frac{\binom{m}{h}\binom{n-m}{k-h}p_k^{(n)}}{\binom{n}{k}p_k^{(n)}} = \frac{\binom{m}{h}\binom{n-m}{k-h}}{\binom{n}{k}} \tag{1.6}$$

By (1.6) and by Bayes' formula, we immediately obtain

$$P\{S_n = k|S_m = h\} = \frac{\binom{m}{h}\binom{n-m}{k-h}}{\binom{n}{k}} \frac{\omega_k^{(n)}}{\omega_h^{(m)}}. \tag{1.7}$$

Furthermore we have

$$P\{S_n = k|X_1, ..., X_m\} = P\{S_n = k|S_1, ..., S_m\} = P\{S_n = k|S_m\} \tag{1.8}$$

and, when $\{E_1, ..., E_n\}$ is a subfamily of a larger family of exchangeable events $\{E_1, ..., E_n, E_{n+1}, ..., E_M\}$,

$$P\{S_m = h|S_n = k, X_{n+1}, ..., X_M\} = P\{S_m = h|S_n = k\} \tag{1.9}$$

Proposition 1.9. *For an arbitrary exchangeable family,*

$$\omega_h^{(m)} = \sum_{k=h}^{n-m+h} \frac{\binom{k}{h}\binom{n-k}{m-h}}{\binom{n}{m}}\omega_k^{(n)}, \ 0 \leq h \leq m. \tag{1.10}$$

Proof. By the rule of total probabilities:

$$\omega_h^{(m)} = \sum_{k=0}^{n} P\{S_m = h|S_n = k\}P\{Z_n = k\}$$

$$= \sum_{k=h}^{n-m+h} P\{S_m = h|Z_n = k\}\omega_k^{(n)}$$

whence Equation (1.10) follows by the formula (1.6). $\qquad\square$

A trivial but fundamental consequence of the definition of exchangeability is illustrated in the following.

Remark 1.10. Let $\mathcal{E} \equiv \{E_1, ..., E_n\}$ be an exchangeable family. For $1 \leq m \leq n$, the joint distribution of $(X_1, ..., X_m)$ is uniquely determined by the probability distribution of the variable S_n.

More precisely, for $(x_1, ..., x_m) \in \{0,1\}^m$ such that $\sum_{i=1}^{m} x_i = h$ ($0 \leq h \leq m$), we can write

$$p_h^{(m)} \equiv P\{X_1 = x_1, ..., X_m = x_m\} = \frac{\omega_h^{(m)}}{\binom{m}{h}}$$

$$= \sum_{k=h}^{n-m+h} \frac{\binom{k}{h}\binom{n-k}{m-h}}{\binom{n}{m}\binom{m}{h}} \omega_k^{(n)} = \sum_{k=h}^{n-m+h} \frac{\binom{n-m}{k-h}}{\binom{n}{k}} \omega_k^{(n)}.$$

Example 1.11. In a lot of n units $\{U_1, ..., U_n\}$, let S_m be the number of defectives out of the sample $\{U_1, ..., U_m\}$. By imposing exchangeability on the events

$$E_1 \equiv \{U_1 \text{ is defective}\}, ..., E_n \equiv \{U_n \text{ is defective}\},$$

we have that the distribution of S_m is determined by that of S_n via Equation (1.10).

If $P\{S_n = k\} = 1$, for some $k \leq n$, then the distribution of S_m is obviously hypergeometric. Generally (1.10) shows that the distribution of S_m is a mixture of hypergeometric distributions.

If S_n is binomial $b(n, p)$ then S_m is binomial $b(m, p)$.

If S_n is uniformly distributed over $\{0, 1, ..., n\}$, then S_m is uniformly distributed over $\{0, 1, ..., m\}$ (see Example 1.19).

Remark 1.12. Let $\omega_h^{(m)}$, $0 \leq h \leq m \leq n$, be the set of probabilities associated with a given exchangeable family. Consider now the scheme of random drawings without replacement from an urn containing M balls, among which M_1 are green as in Example 1.7 (M_1 is a random number).

More precisely, consider the case with $M = n$ and let, for $k = 0, 1, ..., n$,

$$P\{M_1 = k\} = \omega_k^{(n)}.$$

$n = M$ means that "we sample without replacement until the urn is empty").

Thus, for the family, $\widehat{\mathcal{E}} \equiv \{\widehat{E}_1, ..., \widehat{E}_n\}$, with

$$\widehat{E}_i \equiv \{\text{green ball at the i-th drawing}\},$$

we have

$$\widehat{S}_n = \widehat{S}_M = M_1,$$

whence

$$\widehat{\omega}_k^{(n)} \equiv P\{\widehat{S}_n = k\} = P\{M_1 = k\} = \omega_k^{(n)}$$

In view of Proposition 1.9,

$$\omega_h^{(m)} = \widehat{\omega}_h^{(m)}, 0 \leq h \leq m < n.$$

We can then conclude that, for any finite family of exchangeable events, one can find another family, describing drawings without replacement from an urn, sharing the same joint probability distribution. This argument also shows that Equation (1.10) could be obtained by means of a direct heuristic reasoning, taking into account the formula (1.5).

1.2.1   Extendibility and de Finetti's theorem

Any subfamily $\{E_{i_1}, ..., E_{i_m}\}$ of $\mathcal{E}$ is of course still a family of exchangeable events; thus we can wonder if $\mathcal{E}$ is, in its turn, a subfamily of a larger family of exchangeable events.

Definition 1.13. $\mathcal{E} \equiv \{E_1, ..., E_n\}$ is an $N-$extendible family of exchangeable events $(N > n)$ if we can define events $E_{n+1}, ..., E_N$ such that $\{E_1, ..., E_N\}$ are exchangeable.

$\mathcal{E}$ is infinitely extendible if it is N-extendible for any $N > n$.

$\mathcal{E}$ has maximum rank N_0 if $\mathcal{E}$ is N_0-extendible but not $(N_0 + 1)$-extendible.

A necessary and sufficient condition for N-extendibility $(N > n)$ will be given in terms of the quantities $\omega_k^{(n)}$ (Remark 1.15).

Examples 1.4 and 1.5 show cases of infinite extendibility, while in Example 1.7 we have a case of finite extendibility $(N_0 = M)$.

In Example 1.8, $N_0 \geq M$ and the actual value of N_0 depends on the probabilities $P\{M_1 = k\}(k = 0, ..., M)$.

In Example 1.6 we have no given constraint on N_0 (besides, of course, $N_0 \geq n$). However it can be natural to *assume* infinite extendibility.

Remark 1.14. Let $Y_1, ..., Y_n$ be random variables satisfying the conditions:

$$Var(Y_i) = \sigma^2 \ (i = 1, ..., n); \ Cov(Y_i, Y_j) = \rho \cdot \sigma^2$$

(ρ has obviously the meaning of correlation coefficient). By considering the variance of the variable $\sum_{i=1}^{n} Y_i$, we obtain the inequality

$$0 \leq Var\left(\sum_{i=1}^{n} Y_i\right) = \sum_{i=1}^{n} Var\left(Y_i\right) + \sum_{i \neq j} Cov\left(Y_i, Y_j\right) =$$

$$n\sigma^2 + n(n-1)\rho \cdot \sigma^2$$

whence

$$\rho \geq -\frac{1}{n-1}.$$

For the indicators of exchangeable events, the above conditions trivially hold with

$$\sigma^2 = P(E_1) - [P(E_1)]^2 = \omega_1 - (\omega_1)^2$$

and

$$\rho \equiv \frac{P(E_1 \cap E_2) - [P(E_1)]^2}{\sigma^2} = \frac{\omega_2 - (\omega_1)^2}{\omega_1 - (\omega_1)^2}.$$

It then follows that if $\rho < 0$ (i.e. if $(\omega_1)^2 > \omega_2$), we have, for the maximum rank,

$$N_0 < 1 + \frac{1}{|\rho|} = 1 + \frac{\omega_1 - (\omega_1)^2}{(\omega_1)^2 - \omega_2}.$$

Thus $\rho \geq 0$ $((\omega_1)^2 \leq \omega_2)$ is a necessary condition for infinite extendibility.

Remark 1.15. A necessary and sufficient condition for N-extendibility of $\{E_1, ..., E$ $(N > n)$ is the existence of $(N+1)$ quantities $\omega_j^{(N)} \geq 0$ such that

$$\sum_{j=0}^{N} \omega_j^{(N)} = 1, \quad \omega_k^{(n)} = \sum_{j=k}^{N-n+k} \frac{\binom{j}{k}\binom{N-j}{n-k}}{\binom{N}{n}} \omega_j^{(N)}, \quad 0 \leq k \leq n.$$

$\omega_j^{(N)}$ can be interpreted as the probability of j successes out of N trials.

For our purposes it is convenient to rewrite the above condition of N-extendibility in the form

$$\omega_k^{(n)} = \int_0^1 \frac{\binom{\theta N}{k}\binom{(1-\theta)N}{n-k}}{\binom{N}{n}} d\Pi_0^{(N)}(\theta). \tag{1.11}$$

where the symbol $\Pi_0^{(N)}(\theta)$ denotes the probability distribution of the variable

$$\Theta_N \equiv \frac{S_N}{N}$$

(Θ_N is the "frequency of successes among the first N events $E_1, ..., E_N$").

Starting from equation (1.11), we wonder which constraints must be satisfied by the probabilities $\omega_k^{(n)}$ in the case of infinitely extendible families.

This is explained by the following well known result. We shall not prove it formally; rather it can be instructive to sketch two different types of proofs.

Theorem 1.16. *(de Finetti's theorem for exchangeable events). Let $\{E_1, E_2, ...\}$ be a denumerable family of exchangeable events. Then there exists a probability distribution Π_0 on $[0, 1]$ such that*

$$\omega_k^{(n)} = \int_0^1 \binom{n}{k} \theta^k (1 - \theta)^{n-k} d\Pi_0(\theta). \tag{1.12}$$

Proof. 1. We can easily prove the existence of a variable Θ such that

$$\lim_{N\to\infty} |\Theta_N - \Theta|^2 = 0 \ a.s.$$

("law of large numbers for exchangeable events"). This implies that the sequence $\{\Pi_0^{(N)}\}$ converges to a limiting distribution Π_0. On the other hand, as is well known,

$$\lim_{N\to\infty} \frac{\binom{\theta N}{k}\binom{N-\theta N}{n-k}}{\binom{N}{n}} = \theta^k (1 - \theta)^{n-k}$$

the limit being uniform in θ.

We can then obtain (1.12) by letting $N \to \infty$ in Equation (1.11). $\qquad\square$

Proof. 2. Consider the sequence $1 = \omega_0 \geq \omega_1 \geq \omega_2 \geq ...$ and the quantities

$$p_k^{(n)} = \frac{\omega_k^{(n)}}{\binom{n}{k}}.$$

From Equation (1.3) we can in particular obtain

$$p_{n-1}^{(n)} = \omega_n - \omega_{n-1} \equiv -\Delta\omega_{n-1}, \ p_{n-2}^{(n)} = p_{n-2}^{(n-1)} - p_{n-1}^{(n)} \equiv \Delta^2\omega_{n-2}, ... \tag{1.13}$$

This shows that the sequence $\omega_0 \geq \omega_1 \geq \omega_2 \geq ...$ is completely monotone: for $k \leq n$,

$$(-1)^n \Delta^{n-k}\omega_{n-k} \geq 0.$$

Being $\omega_0 = 1$, it follows by Hausdorff theorem (see e.g. the probabilistic proof in Feller, 1971) that there exists a probability distribution Π_0 on $[0, 1]$ such that $\omega_0 \geq \omega_1 \geq \omega_2 \geq ...$ is the sequence of the moments of Π_0:

$$\omega_n = \int_0^1 \theta^n d\Pi_0(\theta). \tag{1.14}$$

Finally (1.12) is obtained by using the representation (1.14) in the sequence of identities (1.13). $\qquad\square$

Remark 1.17. Equation (1.12), $\forall \mathbf{x} \in \{0,1\}^n$ such that $\sum_{i=1}^{n} x_i = k$, can equivalently be rewritten in the form

$$P\{X_1 = x_1, ... X_n = x_n\} = \int_0^1 \theta^k (1-\theta)^{n-k} d\Pi_0(\theta) \qquad (1.15)$$

If Θ is a degenerate random variable, i.e. if $P\{\Theta = \widehat{\theta}\} = 1$ for some $\widehat{\theta} \in (0,1)$, (1.15) becomes

$$P\{X_1 = x_1, ... X_n = x_n\} = \widehat{\theta}^k (1-\widehat{\theta})^{n-k}.$$

Then $E_1, ..., E_n$ are independent and equiprobable. For the general case of infinity extendibility, when Θ is not degenerate, $E_1, ..., E_n$ are conditionally independent and equiprobable, given Θ.

1.2.2 The problem of prediction

Often when dealing with a family of exchangeable events the real object of statistical interest is the *prediction problem,* namely the one of computing, for given $\mathbf{x}' \equiv (x_1', ..., x_m') \in \{0,1\}^m$, the conditional probability

$$P\{X_{m+1} = 1 | X_1 = x_1', ..., X_m = x_m'\},$$

or more generally, for given n and $\mathbf{x} \equiv (x_1, ..., x_n) \in \{0,1\}^n$,

$$P\{X_{m+1} = x_1, ..., X_{m+n} = x_n | X_1 = x_1', ..., X_m = x_m'\}$$

(of course it must be $n + m \leq N_0$). Denoting $h = \sum_{i=1}^{m} x_i'$, we have

$$P\{X_{m+1} = 1 | X_1 = x_1', ..., X_m = x_m'\} =$$

$$\frac{P\{X_1 = x_1', ..., X_m = x_m', X_{m+1} = 1\}}{P\{X_1 = x_1', ..., X_m = x_m'\}} =$$

$$\frac{\omega_{h+1}^{(m+1)} / \binom{m+1}{h+1}}{\omega_h^{(m)} / \binom{m}{h}} = \frac{h+1}{m+1} \frac{\omega_{h+1}^{(m+1)}}{\omega_h^{(m)}}. \qquad (1.16)$$

Similarly, for $(x_1, ..., x_n)$ such that $\sum_{i=1}^{n} x_i = k$, we obtain

$$P\{X_{m+1} = x_1, ... X_{m+n} = x_n | X_1 = x_1', ..., X_m = x_m'\} =$$

$$\frac{\omega_{h+k}^{(n+m)} / \binom{n+m}{h+k}}{\omega_h^{(m)} / \binom{m}{h}}. \qquad (1.17)$$

Example 1.18. Having initially assessed a probability distribution for the number S_n of defective units in a lot of size n, we want to compute the conditional probability,

$$P\{X_{m+1} = 1|S_m = h\},$$

that the next inspected unit will reveal to be defective, given that we found h defective units in the previous m inspections.

To this aim, we apply the formula (1.16). For instance we obtain:

$$P\{X_{m+1} = 1|S_m = h\} = \frac{k-h}{n-m}$$

in the case $P\{S_n = k\} = 1$, for some $k \leq n$,

$$P\{X_{m+1} = 1|S_m = h\} = p$$

in the case when the distribution of S_n is $b(n,p)$, and

$$P\{X_{m+1} = 1|S_m = h\} = \frac{h+1}{n+2}$$

when S_n is uniformly distributed over $\{0, 1, ..., n\}$ (see also Example 1.19).

In the case of an infinitely extendible family we can use the representation (1.12) and then the formulae (1.16), (1.17), respectively, become

$$P\{X_{m+1} = 1|X_1 = x_1', ..., X_m = x_m'\} =$$

$$\frac{\int_0^1 \theta^{h+1}(1-\theta)^{m-h}d\Pi_0(\theta)}{\int_0^1 \theta^h(1-\theta)^{m-h}d\Pi_0(\theta)} \tag{1.18}$$

$$P\{X_{m+1} = x_1, ... X_{m+n} = x_n|X_1 = x_1', ..., X_m = x_m'\} =$$

$$\frac{\int_0^1 \theta^{h+k}(1-\theta)^{n+m-h-k}d\Pi_0(\theta)}{\int_0^1 \theta^h(1-\theta)^{m-h}d\Pi_0(\theta)}. \tag{1.19}$$

For $s > 0$, consider the probability distribution on $[0,1]$ defined by

$$d\Pi_{s,m}(\theta) = \frac{\theta^s(1-\theta)^{m-s}d\Pi_0(\theta)}{\int_0^1 \theta^s(1-\theta)^{m-s}d\Pi_0(\theta)}. \tag{1.20}$$

Equations (1.18) and (1.19) can be rewritten

$$P\{X_{m+1} = 1|X_1 = x_1', ..., X_m = x_m'\} = \int_0^1 \theta d\Pi_{h,m}(\theta) \tag{1.21}$$

$$P\{X_{m+1} = x_1, ...X_{n+m} = x_n | X_1 = x_1', ..., X_m = x_m'\} =$$

$$\int_0^1 \theta^k (1 - \theta)^{n-k} d\Pi_{h,m}(\theta). \tag{1.22}$$

It is of interest to contrast Equations (1.22) and (1.15).

Arguments presented so far can be illustrated by discussing a very simple and well-known case.

Example 1.19. (Inference scheme of Bayes-Laplace). Consider n exchangeable events with a joint distribution characterized by the position

$$\omega_k^{(n)} = \frac{1}{(n+1)}, k = 0, 1, ..., n.$$

Namely, we assess that the number of successes out of the n trials has a uniform distribution over the set of possible values $\{0, 1, ..., n\}$. In this case the formula (1.10) gives, for $0 \le h \le m < n$,

$$\omega_h^{(m)} = \frac{1}{(n+1)\binom{n}{m}} \sum_{k=h}^{n-m+h} \binom{k}{h} \binom{n-k}{m-h}.$$

For $m = n - 1$, one immediately obtains $\omega_h^{(n-1)} = \frac{1}{n}$ and then, by backward induction,

$$\omega_h^{(m)} = \frac{1}{m+1}, \quad 0 \le h \le m \le n - 1.$$

This also shows that we are in a case of infinite extendibility. Indeed, for any $N > n$, we can write, for $\omega_j^{(N)} = \frac{1}{(N+1)}$,

$$\omega_k^{(n)} = \frac{1}{(n+1)} = \sum_{j=k}^{N-n+k} \frac{\binom{j}{k}\binom{N-j}{n-k}}{\binom{N}{n}} \omega_j^{(N)} .$$

Then, by Theorem 1.16, we can write

$$\omega_k^{(n)} = \frac{1}{(n+1)} = \int_0^1 \binom{n}{k} \theta^k (1 - \theta)^{n-k} d\Pi_0(\theta)$$

Π_0 must be the limit of the uniform distribution over $\{0, \frac{1}{N}, \frac{2}{N}, ..., \frac{N-1}{N}, 1\}$, for $N \to \infty$.

Indeed Π_0 is the uniform distribution on the interval $[0, 1]$ and, for $k = 0, 1, ..., n$, we have the well-known identity

$$\frac{1}{(n+1)} = \int_0^1 \binom{n}{k} \theta^k (1 - \theta)^{n-k} d\theta.$$

By applying formula (1.16), or (1.21), we get, for $\mathbf{x}' \in \{0,1\}^m$ such that $\sum_{i=1}^{n} x'_i = h$,

$$P\{X_{m+1} = 1 | X_1 = x'_1, ..., X_m = x'_m\} = \frac{h+1}{m+2}$$

1.2.3   More on infinitely extendible families

We can now present some more comments about infinitely extendible families of exchangeable events.

In the case $N_0 = \infty$, we consider the variable

$$\Theta \equiv \lim_{N \to \infty} \frac{S_N}{N}$$

introduced in the proof 1 of Theorem 1.16.

We recall that the distribution Π_0 appearing in (1.12) has the meaning of "initial" distribution of Θ.

We also note that, for $s = 0, 1, ..., m$, the distribution $\Pi_{s,m}$, defined by Equation (1.20) can be interpreted as the final distribution of Θ after observing $\{S_m = s\}$ (i.e. after observing s successes out of the first m trials).

In the beginning, $\Pi_{S_m,m}$ is a "random distribution", i.e. a distribution which will depend on future results of the first n observations.

However, before taking the observations, one can claim that, for $m \to \infty$, $\Pi_{S_m,m}$ will converge to a degenerate distribution.

As already noticed (Remark 1.17), $E_1, E_2, ...$ are conditionally independent and equiprobable given Θ. Two alternative cases are possible:

a) The initial distribution Π_0 is a degenerate one

b) Π_0 is a non-degenerate distribution.

In case a), $E_1, E_2, ...$ are independent and equiprobable.

The personal probability $P(E_i)$ coincides with the value $\hat{\theta}$ taken by Θ. This is an extreme case, where a person asserts that there is nothing to learn from the observations: $\Pi_{S_m,m}$ trivially coincides with Π_0, with probability one for any m and, as a consequence of (1.21), it is

$$P\{X_{m+1} = 1 | X_1, ..., X_m\} = \hat{\theta} \text{ a.s., } m = 1, 2, ...$$

Case b) is more common. In this case $E_1, E_2, ...$ are not independent but positively correlated. Θ can be interpreted as a parameter.

On the basis of m past observations, the probability of success in a single trial is updated according to formula (1.21).

When m is very large, such a probability is approximated by the observed frequency of successes.

The above discussion can be reformulated as follows:

Remark 1.20. Let $E_1, E_2, \ldots$ be independent events with $P(E_i) = \widehat{\theta}$ ($i = 1, 2, \ldots$), i.e. they are exchangeable with

$$\omega_m = (\widehat{\theta})^m, \quad m = 1, 2, \ldots \tag{1.23}$$

or, equivalently,

$$\omega_h^{(m)} = \binom{m}{h}(\widehat{\theta})^m(1 - \widehat{\theta})^{m-h}, 0 \leq h \leq m, m = 1, 2, \ldots.$$

Under this condition, by the law of large numbers, it is

$$\frac{S_N}{N} \to \widehat{\theta}, \text{ a.s.} \tag{1.24}$$

i.e. $\Theta = \lim_{N \to \infty} \frac{S_N}{N}$ is a degenerate random variable. On the other hand, for a denumerable sequence of exchangeable events, the validity of the limit (1.24), for some given $\widehat{\theta} \in (0, 1)$, is not only a necessary condition, but also a sufficient condition for the condition (1.23) to hold.

If, on the contrary, the initial distribution Π_0 of Θ (i.e. the limiting distribution of $\frac{S_N}{N}$) is not degenerate, the events are positively correlated.

In both cases, however, it is

$$P\{X_{N+1} = 1 | X_1, \ldots, X_N\} - \frac{S_N}{N} \to 0, \text{ a.s.}$$

or

$$P\{X_{N+1} = 1 | X_1, \ldots, X_N\} \to \Theta, \text{ a.s.}$$

Remark 1.21. By means of the concept of an infinitely extendible family we give a mathematical meaning to the distributions Π_0 and $\Pi_{s,n}$ which are fundamental in the practice of Bayesian statistics. Moreover we recover the concepts of the frequentist approach in subjectivist terms. $\Theta \equiv \lim_{N \to \infty} \frac{S_N}{N}$ can be seen as the "frequentist probability". Frequentist probability and personal probability are two different concepts, anyway. They only occasionally coincide in the case of sequences of events which are (in the subjective point of view) independent and equiprobable.

We see from (1.12) that infinitely extendible families can be represented by models of random drawings with replacements from urns (we already saw that an arbitrary family can be represented by models of drawings without replacement).

Case a) corresponds to urns with known composition, while case b) corresponds to urns with unknown composition.

Remark 1.22. It is to be stressed however that in most cases the urn is purely *hypothetical* and one thinks of that only for the sake of conceptual simplicity. This happens in all cases when we are not dealing with a real finite population from which we draw with replacement. In these cases the parameter Θ has a purely hypothetical meaning and it is only introduced for mathematical convenience.

To clarify the above, compare the situations described in Examples 1.5 and 1.6, respectively.

In the former case the aim of statistical analysis is typically the estimation of the quantity Θ, since we draw observations from a real finite population and Θ has a clear physical meaning of its own.

In the latter case the object of statistical interest is rather the prediction problem. However it can be natural, as already mentioned, to assume infinite extendibility and then to define the variable Θ, also for the present case. Θ has the meaning of being the limit of

$$P\{X_{N+1} = 1 | X_1, ..., X_N\},$$

for $N \to \infty$.

We remark, anyway, that the concepts of subjective probability and exchangeability put us in a position of dealing with the broader class of inference problems which only concern with finite collections of similar events. In such problems the inference problem has a more realistic meaning and prediction is to be made about the quantity

$$S_{N_0} = \sum_{i=1}^{N_0} X_i.$$

We conclude this section by presenting a classical model of "contagious probabilities". This provides an example of conditional independence, arising from an urn model different from simple random sampling with replacement.

Example 1.23. (Polya urns and Bayes-Laplace model). An urn initially contains M_1 green balls and $M - M_1$ red ones, with M and M_1 known quantities, like in the Examples 1.4 and 1.7.

We perform random drawings from the urn but, this time, after any drawing we put the sampled ball again into the urn together with one additional ball of the same color. Denote again

$$E_i \equiv \{ \text{ a green ball at the i-th drawing}\}, \ i = 1, 2, ...$$

This gives rise to an infinitely extendible family of exchangeable events.

In the present case, we have a special type of positive dependence among events due to the direct influence of observed results on the probability of success in the next trial.

More precisely, it obviously is, for $(x_1', ..., x_m') \in \{0,1\}^m$ such that $\sum_{i=1}^{m} x_i' = h$,

$$P\{X_{m+1} = 1 | X_1 = x_1', ..., X_m = x_m'\} = \frac{M_1 + h}{M + m}.$$

Noting also that $P\{X_1 = 1\} = \frac{M_1}{M}$, we obtain, for $\mathbf{x} \equiv (x_1, ..., x_n)$ such that $\sum_{i=1}^{n} x_i = k$

$$P\{\mathbf{X} = \mathbf{x}\} = \frac{\omega_k^{(n)}}{\binom{n}{k}} =$$

$$\frac{M_1(M_1 + 1)...(M_1 + k - 1)(M - M_1)...(M - M_1 + n - k - 1)}{M(M + 1)...(M + n - 1)}.$$

Since we deal with an infinitely extendible exchangeable family, there must exist a mixing distribution Π_0 allowing the quantities $\omega_k^{(n)}$ to have the representation (1.12).

Consider in particular the case $M = 2$, $M_1 = 1$. This entails

$$\omega_k^{(n)} = \frac{1}{(n+1)}, \quad k = 0, 1, ..., n.$$

Thus we see that this model coincides with that of Bayes-Laplace discussed in Example 1.19. Then, in this case, Π_0 is the uniform distribution on the interval $[0, 1]$.

1.3 Exchangeable random quantities

In the previous section we dealt with binary random quantities; this allowed us to describe states of probabilistic indifference among several random events. In the present section we aim to formalize the condition of indifference among different random quantities; to this purpose we extend the concept of exchangeability to the case of real random variables, $X_1, X_2, ...$.

Example 1.24. Consider the lot of n units as in Example 1.1. Here we do not classify units according to being defective or good; rather, the quantity X_i which we associate to the unit U_i $(i = 1, ..., n)$ is not binary; X_i might be for instance the life-length of U_i or its resistance to a stress, and so on.

We want to provide a probabilistic description of the case when, before testing, there is no reason to distinguish among the units, as far as our own opinion about $X_1, ..., X_n$ is concerned.

The situation of indifference relative to $X_1, ..., X_n$ in the example above is formalized by means of the condition of exchangeability for $X_1, ..., X_n$.

Definition 1.25. $X_1, ..., X_n$ are exchangeable if they have the same joint distribution as $X_{j_1}, ..., X_{j_n}$, for any permutation $\{j_1, ..., j_n\}$ of the indexes $\{1, 2, ..., n\}$.

In other words $X_1, ..., X_n$ are exchangeable if the joint distribution function

$$F^{(n)}(x_1, ..., x_n) \equiv P\{X_1 \leq x_1, ..., X_n \leq x_n\}$$

is a symmetric function of its arguments $x_1, ..., x_n$. Such a condition then means that the state of information on $(X_1, ..., X_n)$ is symmetric.

We note that of course exchangeability is actually a property of a joint distribution function $F^{(n)}(x_1, ..., x_n)$; when convenient we speak of an exchangeable distribution function in place of a symmetric distribution function.

Remark 1.26. Independent, identically distributed, random variables are trivially exchangeable.

Most of the time we shall be dealing with the case of *absolutely continuous* distributions, for which

$$F^{(n)}(x_1, ..., x_n) = \int_{-\infty}^{x_1} \int_{-\infty}^{x_2} \cdots \int_{-\infty}^{x_n} f^{(n)}(\xi_1, ..., \xi_n) d\xi_1 ... d\xi_n$$

where $f^{(n)}$ denotes the joint probability density function.

In such cases it is

$$f^{(n)}(x_1, ..., x_n) = \frac{\partial^n}{\partial x_1 ... \partial x_n} F_n(x_1, ..., x_n)$$

and, as it is easy to see, $X_1, ..., X_n$ are exchangeable if and only if $f^{(n)}$ is symmetric as well.

Example 1.27. Let (X_1, X_2) be a pair of real random variables, uniformly distributed over a region B of the plane:

$$f^{(2)}(x_1, x_2) = \begin{cases} \frac{1}{\text{area}(B)} & \text{if } (x_1, x_2) \in B \\ 0 & otherwise \end{cases} .$$

Then (X_1, X_2) are exchangeable if and only if B is symmetric with respect to the line $x_2 = x_1$.

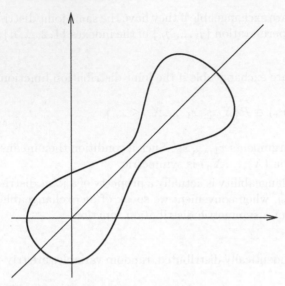

Figure 1 Example of a symmetric set

(X_1, X_2) are i.i.d. if and only if B is of the type:

$$B \equiv \{(x_1, x_2) \,|\, x_1 \in A, x_2 \in A\}$$

for some (measurable) region A of the real line.

Several further examples of exchangeable random variables will be discussed in the next chapter; examples there are however restricted to the case of non-negative variables.

The concept of exchangeable random variables was introduced by de Finetti (1937). It characterizes, in a setting of subjective probability, the common concept of "random sample from a population". The distinction between sampling from finite and infinite populations is rendered in terms of the notions of finite and infinite extendibility, which we are going to recall next.

1.3.1 Extendibility and de Finetti's theorem for exchangeable random variables

We note that if $X_1, ..., X_n$ are exchangeable, $X_{j_1}, ..., X_{j_m}$ also are exchangeable, with $\{j_1, ..., j_m\}$ any subset of $\{1, 2, ..., n\}$. The joint distribution function of $X_{j_1}, ..., X_{j_m}$,

$$F^{(m)}(x_1, ..., x_m) \equiv \lim_{x_{m+1}, ..., x_n \to \infty} F^{(n)}(x_1, ..., x_n)$$

depends of course on m and on $F^{(n)}$ but it does not depend on the particular choice of the subset $\{j_1, ..., j_m\}$.

Definition 1.28. An exchangeable distribution function $F^{(n)}(x_1, ..., x_n)$ is N-*extendible* $(N > n)$ if we can find an exchangeable N-dimensional distribution function $F^{(N)}$ such that $F^{(n)}$ is a (n-dimensional) marginal distribution of $F^{(N)}$.

$F^{(n)}$ is *infinitely extendible* if it is N-extendible for any $N > n$.

For a given exchangeable distribution function $F^{(n)}(x_1, ..., x_n)$, the m-dimensional marginal $F^{(m)}(x_1, ..., x_m)$ is of course n-extendible.

Remark 1.29. If $X_1, ..., X_n$ are such that

$$\rho = \frac{Cov(X_i, X_j)}{Var(X_i)} < 0, (1 \leq i \neq j \leq n)$$

then their distribution cannot be N-extendible for $N > 1 + \frac{1}{|\rho|}$ (see Remark 1.14).

Example 1.30. Consider n exchangeable random quantities $X_1, ..., X_n$ such that, for some deterministic value K, it is

$$P\{\sum_{i=1}^{n} X_i = K\} = 1 \qquad (1.25)$$

Suppose we could find a further random quantity X_{n+1} such that $X_1, ..., X_n, X_{n+1}$ were exchangeable. By symmetry, $P\{\sum_{i=2}^{n+1} X_i = K\} = 1$ would result, whence

$$P\{X_1 = X_2 = ... = X_n = X_{n+1}\} = 1,$$

i.e. an exchangeable distribution satisfying (1.25) and such that $P\{X_1 = X_2 = ... = X_n\} < 1$ cannot be extendible.

Further examples of non-extendibility arising from applications of our interest will be presented in the next Chapter.

Remark 1.31. From Remark 1.29, it follows that infinite extendibility implies $\rho(X_i, X_j) \geq 0$. This is not a sufficient condition however: we can find positively correlated exchangeable quantities which are not conditionally independent, identically distributed (see e.g. Exercise 1.60).

Our aim now is to discuss how de Finetti's Theorem 1.16 can be extended to the case of infinitely extendible vectors of real exchangeable random variables. We start by considering two situations which obviously give rise to infinite extendibility.

Let $X_1, ..., X_n$ be independent, identically distributed (i.i.d.) and let $G(x)$ denote their common one-dimensional distribution function. The joint distribution function

$$F^{(n)}(x_1, ..., x_n) = G(x_1) \cdot G(x_2) \cdot ... \cdot G(x_n) \qquad (1.26)$$

is obviously infinitely extendible: for any $N > n$, $F^{(n)}$ is the marginal of

$$F^{(N)}(x_1, ..., x_N) = G(x_1) \cdot G(x_2) \cdot ... \cdot G(x_N).$$

Let Θ be a random variable taking values in a space L and with a distribution Π_0. Let moreover $X_1, ..., X_n$ be conditionally i.i.d., given Θ:

$$F^{(n)}(x_1, ..., x_n | \Theta = \theta) = G(x_1 | \theta) \cdot G(x_2 | \theta) \cdot ... \cdot G(x_n | \theta) \qquad (1.27)$$

where $G(x | \theta)$ denotes their common one-dimensional conditional distribution function. The joint distribution function of $X_1, ..., X_n$

$$F^{(n)}(x_1, ..., x_n) = \int_L G(x_1 | \theta) \cdot G(x_2 | \theta) \cdot ... \cdot G(x_n | \theta) d\Pi_0(\theta) \qquad (1.28)$$

is symmetric; furthermore it is obviously infinitely extendible: for any $N > n$, $F^{(n)}$ is the marginal of

$$F^{(N)}(x_1, ..., x_N) = \int_L G(x_1 | \theta) \cdot G(x_2 | \theta) \cdot ... \cdot G(x_N | \theta) d\Pi_0(\theta).$$

We can now illustrate the de Finetti's representation result.

Under quite general conditions, the latter shows that all possible situations of infinite extendibility are of the type (1.28) or of the type (1.26).

In an informal way it can thus be stated as follows:

Theorem 1.32. *The distribution of exchangeable random variables $X_1, ..., X_n$ is infinitely extendible if and only if $X_1, ..., X_n$ are i.i.d. or conditionally i.i.d..*

This result can be formulated, and proven, in a number of different ways. Complete proofs can be found in (de Finetti, 1937; Hewitt and Savage, 1955; Loeve, 1960 and Chow and Teicher, 1978). A much more complete list of references on exchangeability is contained in the exhaustive monograph by Aldous (1983).

The essence of the de Finetti's representation result is based on the special relation existing between a symmetric distribution function $F^{(N)}$ and its n-dimensional marginal $F^{(n)}$ $(n < N)$. Below, we are going to explain this, rather than dwelling on a formal mathematical treatment.

Our discussion will start with some notations and remarks.

First of all for a fixed set of N real numbers $z_1, ..., z_N$ (not necessarily all distinct) denote by

$$\phi_N(\mathbf{z}) \equiv (z_{(1)}, ..., z_{(N)})$$

the vector of the corresponding order statistics. Moreover, denote by the symbol $F_{\mathbf{z}}^{(N)}$ the distribution function corresponding to the discrete N-dimensional probability distribution $P_{\mathbf{z}}^{(N)}$, uniform on the set of permutations of $\{z_1, ..., z_N\}$ and let $F_{\mathbf{z}}^{(N,n)}$ be the distribution function of the discrete n-dimensional probability distribution $P_{\mathbf{z}}^{(N,n)}$, uniform on the set of all the n-dimensional vectors of the form $(z_{j_1}, ..., z_{j_n})$, with $\{j_1, ..., j_n\}$ any subset of $\{1, ..., N\}$.

Remark 1.33. For our purposes, it is useful to look at $P_{\mathbf{z}}^{(N)}$ as the probability distribution of $(Y_1, ..., Y_N)$, where $(Y_1, ..., Y_N)$ is a random permutation of $(z_1, ..., z_N)$, i.e. it is a sample of size N without replacement from a N-size population with elements $\{z_1, ..., z_N\}$.

Similarly, $P_{\mathbf{z}}^{(N,n)}$ can be seen as the probability distribution of a random sample of size n, $(Y_1, ..., Y_n)$, and this shows that $P_{\mathbf{z}}^{(N,n)}$ does coincide with the n-dimensional marginal of $P_{\mathbf{z}}^{(N)}$.

We need to emphasize that $F_{\mathbf{z}}^{(N)}$ and $F_{\mathbf{z}}^{(N,n)}$ depend on $\mathbf{z} = (z_1, ..., z_N)$ symmetrically or, in other words, that they depend on $\mathbf{z}$ only through the vector $\phi_N(\mathbf{z})$; to this aim we shall also write

$$F_{\mathbf{z}}^{(N)} \equiv F_{\phi_N(\mathbf{z})}^{(N)}, \quad F_{\mathbf{z}}^{(N,n)} \equiv F_{\phi_N(\mathbf{z})}^{(N,n)}.$$

For $n = 1$, it is

$$F_{\phi_N(\mathbf{z})}^{(N,1)}(x) = \frac{1}{N} \sum_{i=1}^{N} 1_{[x \le z_i]}, \quad \forall x \in \mathbb{R} \tag{1.29}$$

i.e. $F_{\phi_N(\mathbf{z})}^{(N,1)}(x)$ does coincide with the empirical distribution function which counts, for any real number x, the number of components of $\mathbf{z}$ which are not greater than x.

Remark 1.34. The well known fact that, when the size of a sampled population goes to infinity, random sampling without replacement tends to coincide with random sampling with replacement can be rephrased by using the notation introduced above.

Let $\{z_1, z_2, ...\}$ be a denumerable sequence of real numbers and $\xi \equiv (\xi_1, ..., \xi_n)$ a given vector; then, for $N \to \infty$,

$$F_{\phi_N(\mathbf{z})}^{(N,n)}(\xi_1, ..., \xi_n) \to F_{\phi_N(\mathbf{z})}^{(N,1)}(\xi_1) \cdot .... \cdot F_{\phi_N(\mathbf{z})}^{(N,1)}(\xi_n) \tag{1.30}$$

The primary role of the distributions $P_{\mathbf{z}}^{(N)}$ and $P_{\mathbf{z}}^{(N,n)}$ is explained by the following result. Let $F^{(N)}$ be an N-dimensional symmetric distribution function and let $F^{(n)}$ be its n-dimensional marginal.

Proposition 1.35. $F^{(n)}$ *can be written as a mixture of distribution functions of the type* $F_{\mathbf{z}}^{(N,n)}$, *the mixing distribution being* $F^{(N)}$, *i.e.*

$$F^{(n)}(\xi_1, ..., \xi_n) = \int_{\mathbb{R}^N} F_z^{(N,n)}(\xi_1, ..., \xi_n) dF^{(N)}(\mathbf{z}) \qquad (1.31)$$

Proof. Consider N exchangeable random quantities $Z_1, ..., Z_N$ with joint distribution function $F^{(N)}$.

The property of symmetry means that $F^{(N)}$ coincides with the distribution function of any permutation of $Z_1, ..., Z_N$; i.e., using the notation introduced above

$$F^{(N)}(\xi_1, ..., \xi_N) = \int_{\mathbb{R}^N} F_{\mathbf{z}}^{(N)}(\xi_1, ..., \xi_N) dF^{(N)}(\mathbf{z}) \qquad (1.32)$$

Equation (1.31) can then be immediately obtained by passing to the n-dimensional marginals of both sides of such identity and then taking into account that $F_{\mathbf{z}}^{(N,n)}$ is the n-dimensional marginal of $F_{\mathbf{z}}^{(N)}$:

$$F^{(n)}(\xi_1, ..., \xi_n) = \lim_{\xi_{n+1}, ..., \xi_N \to \infty} F^{(N)}(\xi_1, ..., \xi_N)$$

$$= \int_{\mathbb{R}^N} [\lim_{\xi_{n+1}, ..., \xi_N \to \infty} F_z^{(N)}(\xi_1, ..., \xi_N)] dF^{(N)}(\mathbf{z})$$

$$= \int_{R^N} F_{\mathbf{z}}^{(N,n)}(\xi_1, ..., \xi_N) dF^{(N)}(\mathbf{z})$$

$\square$

We can say that the formula (1.32) is of fundamental importance.

The above arguments can be summarized as follows: exchangeable random variables $X_1, ..., X_N$ have the same distribution of a random permutation of them and this entails that, for $n < N$, $(X_1, ..., X_n)$ has the same joint distribution as a random sample $(Y_1, ..., Y_n)$ from a population formed by the elements $\{X_1, ..., X_N\}$.

The representation (1.31) can be seen as the analog of formula (1.11); for our purposes it is convenient to rewrite it in the form

$$F^{(n)}(\xi_1, ..., \xi_n) = E\left[F_{\phi_N(X_1, ..., X_N)}^{(N,n)}(\xi_1, ..., \xi_n)\right]. \qquad (1.33)$$

The essence of de Finetti's theorem can now be heuristically grasped by letting N go to infinity in (1.33).

First of all consider, $\forall \xi \in \mathbb{R}$, the sequence of random variables

$$\{F_{\phi_N(X_1, ..., X_N)}^{(N,1)}(\xi)\}_{N=1,2,...}$$

By (1.29), $F^{(N,1)}_{\phi_N(X_1,...,X_N)}(\xi)$ is the "frequency of successes" in the family of exchangeable events $\{E_1(\xi), ..., E_N(\xi)\}$, where

$$E_j(\xi) \equiv (X_j \leq \xi).$$

As mentioned in the previous section then, there exists a random quantity $\mathcal{F}(\xi)$ such that $\{F^{(N,1)}_{\phi_N(X_1,...,X_N)}(\xi)\}$ converges to $\mathcal{F}(\xi)$, a.s..

Taking into account that, on the other hand, the limit in (1.30) holds, for $N \to \infty$, we have

$$F^{(N,n)}_{\phi_N(X_1,...,X_N)}(\xi_1, .., \xi_n) \to F^{(N,1)}_{\phi_N(X_1,...,X_N)}(\xi_1) \cdot ... \cdot F^{(N,1)}_{\phi_N(X_1,...,X_N)}(\xi_n)$$

$$\to \mathcal{F}(\xi_1) \cdot ... \cdot \mathcal{F}(\xi_n)$$

Thus, by letting $N \to \infty$ in the equation (1.32), we can write

$$F^{(n)}(\xi_1, ..., \xi_n) = E\left[\mathcal{F}(\xi_1) \cdot ... \cdot \mathcal{F}(\xi_n)\right]. \tag{1.34}$$

$\mathcal{F}(\cdot)$ can be seen as a random one-dimensional distribution function depending on the whole random sequence $\{\phi_N(X_1, X_2, ...X_N)\}$.

We can conclude with the following.

Proposition 1.36. $X_1, X_2, ...$ *are conditionally independent, identically distributed, given* $\mathcal{F}(\cdot)$.

Remark 1.37. The sequence $\{\phi_N(X_1, X_2, ..., X_N)\}$ had, in the last result, a role analogous to the one of the sequence $\{\Theta_N\}$ in the proof of de Finetti's theorem for exchangeable events and $\mathcal{F}(\cdot)$ has now a role analogous to that of the limit random variable Θ there.

Some of the arguments contained in Remark 1.20 can be extended to the more general case.

Remark 1.38. $\mathcal{F}(\cdot)$ can of course be defined only under the condition of infinite extendibility. We stress that $\mathcal{F}(\cdot)$ is in general random (i.e. unknown to us, before taking the observations $X_1, X_2, ...$).

Let us denote by the symbol $\mathcal{L}$ the law of $\mathcal{F}(\cdot)$.

$\mathcal{L}$ is a probability distribution on the space of all one-dimensional distribution functions. $\mathcal{L}$ is a degenerate law concentrated on a given distribution function $\widehat{F}(\cdot)$ if and only if $X_1, X_2, ...$ are independent, identically distributed and $\widehat{F}(\cdot)$ is their one-dimensional distribution. Indeed if $\mathcal{L}$ is degenerate then $X_1, X_2, ...$ are independent by Proposition 1.36. On the other hand if $X_1, X_2, ...$ are independent, identically distributed with a distribution $\widehat{F}(\cdot)$ then the Glivenko-Cantelli theorem (see e.g. Chow and Teicher, 1978) ensures that

$$\sup_{-\infty < x < \infty} |F^{(N,1)}_{\phi_N(X_1,...,X_N)}(\xi) - \widehat{F}(\xi)| \overset{a.c.}{\to} 0$$

whence we can claim that $\mathcal{L}$ is concentrated on $\widehat{F}(\xi)$.

It will be shown in the next section that, under the condition of infinite extendibility, it is in general possible to find different objects, with respect to which $X_1, X_2...$ are conditionally i.i.d..

In any case the "limiting empirical distribution" $\mathcal{F}(\cdot)$ is one of those, as we have just seen.

In special cases (*parametric models*), such "objects" can also be given a finite-dimensional representation; an instance of that is of course provided by denumerable sequences of $\{0, 1\}$-valued exchangeable random quantities.

Parametric models are the common models considered in parametric statistics; they are characterized by the existence of a random variable Θ, taking values in a domain $L \subset \mathbb{R}^k$ (for some $k = 1, 2, ...$) such that $X_1, X_2, ...$are conditionally independent, identically distributed given Θ.

This means that, for a suitable family of one-dimensional distribution functions $\{G(\cdot|\theta)\}_{\theta \in L}$ and a probability distribution Π_0, we have

$$F^{(n)}(x_1, x_2, ..., x_n) = \int_L G(x_1|\theta)G(x_2|\theta)...G(x_n|\theta)d\Pi_0(\theta). \qquad (1.35)$$

In some of the above cases, the parameter Θ has a clear physical meaning of its own.

In other cases the meaning of Θ can be explained in terms of the concept of *predictive sufficiency* and of *de Finetti-type results*, which will be briefly illustrated in the next section.

1.3.2 The problem of prediction

We now turn to considering the problem of *predictive inference* for exchangeable random variables, namely to studying conditional distributions of variables $X_{m+1}, ..., X_{m+n}$, given the values taken by previous observations $X_1, ..., X_m$.

We look at such a conditional distribution as a tool for describing changes of uncertainty, caused by previous observations, about (random) quantities still to be observed.

Of course the conditional distribution is to be obtained from the assessment of the joint distribution of $X_1, ..., X_m, X_{m+1}, ..., X_{m+n}$.

The possibility of dealing with joint and conditional distributions for the variables $X_1, X_2...$, avoiding the intervention of a parameter Θ, is one of the fundamental features of the Bayesian approach.

Here we consider a set of exchangeable random quantities $X_1, ..., X_N$.

We shall assume the existence of a joint density; for $\{j_1, ..., j_k\} \subset \{1, ..., N\}$, the joint density function of k quantities $X_{j_1}, ..., X_{j_k}$ will be denoted by the symbol $f^{(k)}$. Of course, one has

$$f^{(k)}(x_1, ..., x_k) = \int_{-\infty}^{\infty} ... \int_{-\infty}^{\infty} f^{(N)}(x_1, ..., x_k, \xi_1, ..., \xi_{N-k}) \, d\xi_1...d\xi_{N-k} \quad (1.36)$$

For $1 \leq m < n + m \leq N$, the conditional density of $X_{m+1}, ..., X_{m+n}$, given an observation

$$\{X_1 = x'_1, ..., X_m = x'_m\}$$

such that $f^{(m)}(x'_1, ..., x'_m) > 0$, is

$$f^{(n)}(x_{m+1}, ..., x_{m+n}|x'_1, ..., x'_m) =$$

$$\frac{f^{(m+n)}(x'_1, ..., x'_m, x_{m+1}, ..., x_{m+n})}{\int_{-\infty}^{\infty} \cdots \int_{-\infty}^{\infty} f^{(m+n)}(x'_1, ..., x'_m, \xi_{m+1}, ..., \xi_{m+n}) \, d\xi_{m+1}...d\xi_{m+n}} \qquad (1.37)$$

Consider now in particular the special case of parametric models, defined by a joint distribution of the form (1.35), where $X_1, ..., X_n, ...$ are conditionally i.i.d. with respect to the parameter Θ.

In such a case we have absolute continuity if and only if the one-dimensional distributions $G(\cdot|\theta)$ admit density functions $g(\cdot|\theta)$, $\theta \in L$ and the n-dimensional (predictive) density function $f^{(n)}(x_1, ..., x_n)$ takes the form

$$f^{(n)}(x_1, ..., x_n) = \int_L g(x_1|\theta)...g(x_n|\theta) d\Pi_0 \qquad (1.38)$$

For $x'_1, ..., x'_m$ such that

$$f^{(m)}(x'_1, ..., x'_m) > 0, \qquad (1.39)$$

let $\Pi_m(\cdot|\mathbf{x}')$ be the conditional (i.e. "posterior") distribution of Θ, given $\{X_1 = x'_1, ..., X_m = x'_m\}$. By Bayes' theorem, it is

$$d\Pi_m(\cdot|\mathbf{x}') = \frac{g(x'_1|\theta)...g(x'_m|\theta) d\Pi_0}{\int_L g(x'_1|\theta)...g(x'_m|\theta) d\Pi_0} \qquad (1.40)$$

In terms of such a conditional distribution, we can write the conditional density $f^{(n)}(x_{m+1}, ..., x_{m+n}|\mathbf{x}')$ in a shape formally similar to that of equation (1.38) itself, as the following Proposition shows.

Proposition 1.39. *Under the condition (1.38), we have, for $\mathbf{x}' \equiv (x'_1, ..., x'_m)$ such that (1.39) is satisfied,*

$$f^{(n)}(x_{m+1}, ..., x_{m+n}|x'_1, ..., x'_m) =$$

$$\int_L g(x_{m+1}|\theta)...g(x_{m+n}|\theta) d\Pi_m(\theta|\mathbf{x}') \qquad (1.41)$$

Proof. We only must notice that, under the condition (1.38), the formula (1.37) results in

$$f^{(n)}(x_{m+1}, ..., x_{m+n}|x'_1, ..., x'_m) =$$

$$\frac{\int_L g(x_{m+1}|\theta)...g(x_{m+n}|\theta)g(x'_1|\theta)...g(x'_m|\theta)d\Pi_0}{\int_L g(x'_1|\theta)...g(x'_m|\theta)d\Pi_0}$$

whence the equation (1.41) immediately follows by taking into account the identity (1.40). □

Note that the equation (1.41) has a clear heuristic interpretation in terms of conditional independence between $(X_1, ..., X_m)$ and $(X_{m+1}, ..., X_{m+n})$, given Θ.

The following Remark aims to illustrate heuristically the meaning of the distribution Π_0, in a parametric model. A complete treatment is beyond the scope of this monograph. However the arguments below can be seen as an appropriate extension of those presented in Remark 1.20.

Remark 1.40. Π_0 is degenerate if and only if $X_1, X_2, ...,$ are i.i.d.. In such a case also

$$\Pi_m(\cdot|X_1, ..., X_m)$$

is obviously degenerate, with probability one, for any m.

In other cases one can claim that $\Pi_m(\cdot|X_1, ..., X_m)$ will converge, with probability one, toward a degenerate distribution, concentrated on some value

$$\widehat{\theta}(X_1, X_2, ...).$$

$\widehat{\theta}(X_1, X_2, ...)$ is a L-valued random variable, which has the meaning of *Bayes asymptotic estimate* of Θ and, in such cases, its distribution does coincide with Π_0.

It is important to notice the following consequence of equation (1.41) and of the fact that $\Pi_m(\cdot|X_1, ..., X_m)$ is, asymptotically, degenerate: conditionally on a very large number N of past observations $X_1, ..., X_N$, further observations $X_{N+1}, ..., X_{N+n}$, tend to become independent, identically distributed.

Remark 1.41. Think of the situation in which $X_1, X_2, ...$ are i.i.d. with a distribution depending on a parameter Θ; $X_1, X_2, ...$ are trivially exchangeable.

The property of independence is lost when Θ is considered as a random variable with a distribution Π_0 of its own and we "uncondition" with respect to Θ, i.e. when we pass from a joint distribution of the form (1.27) to one of the form (1.28).

By contrast the property of exchangeability is maintained under "unconditioning".

Remark 1.42. Stochastic independence is, from a Bayesian standpoint, a stronger condition than physical independence: in most cases we deal with random quantities representing physically independent phenomena, but such that, due to the common dependence of their distribution on the same (random) factor, are not stochastically independent. This aspect will be analyzed further in Subsection 3.3. of Chapter 3.

Finally we want to direct the reader's attention toward a further class of cases in which changes in the state of information destroy independence without affecting exchangeability.

Remark 1.43. Let $X_1, ..., X_n$ be i.i.d. and let $Z(x_1, ..., x_n)$ be a symmetric function with values in the image space $Z(\mathbb{R}^n)$. Fix a subset $\mathcal{S}_1$ of $Z(\mathbb{R}^n)$ and consider the conditional distribution of $X_1, ..., X_n$, given the event

$$\{Z(X_1, ..., X_n) \in \mathcal{S}_1\}.$$

Under such a conditioning operation, the property of independence is generally lost. However, due to the symmetry of Z, exchangeability is maintained.

We note that, in such cases, we in general meet situations of finite exchangeability.

1.4   de Finetti-type theorems and parametric models

The classic de Finetti's Theorem 1.32 shows that the assumption of exchangeability for a denumerable sequence of random quantities is equivalent to a model of (conditional) independence and identical distribution.

The law of the random distribution $\mathcal{F}$, appearing in the de Finetti's representation, is in general a probability law $\mathcal{L}$ on the space of all the one-dimensional distribution functions (see Remark 1.38).

As we saw, assessing that $\mathcal{L}$ is a degenerate law concentrated on a single distribution function $\widehat{F}$ is equivalent to impose that $X_1, X_2, ...$ are, in particular, i.i.d. (with distribution $\widehat{F}$).

We can claim that a parametric model of the form (1.35) arises when we assess that $\mathcal{L}$ is concentrated on a parametrized sub-family $\{G(\cdot|\theta)\}_{\theta \in L}$.

Of course parametric models are very important both in the theory and in the applications. We shall see, in particular, that notions to be developed in the next chapters take special forms when probabilistic assessment is described by means of parametric models.

For this reason it is of interest to show possible motivations for assessments of this special kind.

Motivations can come from symmetry (or invariance) assumptions, more specific than simple exchangeability, as is demonstrated by de Finetti-type results.

A de Finetti-type theorem for a sequence of exchangeable random quantities $X_1, X_2, ...$ is any result proving that a specific situation of conditional independence, with a fixed family of conditional distributions, is characterized by means of certain invariance properties of n-dimensional ($n = 1, 2, ...$) predictive distributions.

This will be illustrated by concentrating attention on a particular case, which has special interest in our setting.

We assume existence of joint density functions and the symbol $f^{(n)}$ will again denote the n-dimensional joint density of n variables $X_{j_1}, X_{j_2}, ...X_{j_n}$.

Theorem 1.44. *Let $X_1, X_2, ...$ be non-negative random variables and let the density functions $f^{(n)}$ ($n = 2, 3, ...$) satisfy the following implication:*

$$\sum_{i=1}^{n} x_i' = \sum_{i=1}^{n} x_i'' \Rightarrow f^{(n)}(x_1', ..., x_n') = f^{(n)}(x_1'', ..., x_n''). \tag{1.42}$$

Then, for a suitable probability distribution $\Pi_0(\theta)$ on $[0, +\infty)$,

$$f^{(n)}(x_1, ..., x_n) = \int_0^\infty \theta^n \exp\{-\theta \sum_{i=1}^{n} x_i\} d\Pi_0(\theta) \tag{1.43}$$

Proof. Put

$$\overline{F}^{(1)}(t) \equiv P\{X_1 > t\}; \quad \phi_1(t) \equiv f^{(1)}(t) = -\frac{d}{dt}\overline{F}^{(1)}(t), t \geq 0.$$

By the assumptions (1.42), we can find non-negative functions $\phi_2, \phi_3, ...$ such that

$$f^{(n)}(x_1, ..., x_n) = \phi_n(\sum_{i=1}^{n} x_i). \tag{1.44}$$

By combining the equation (1.44) with the condition that $f^{(n)}$ is the n-dimensional marginal of the density $f^{(n+1)}$, we obtain, for $n = 1, 2, ...,$

$$\phi_n(\sum_{i=1}^{n} x_i) = \int_0^\infty \phi_{n+1}(\sum_{i=1}^{n} x_i + u)du,$$

whence

$$\phi_n(t) = \int_t^\infty \phi_{n+1}(v)dv.$$

This shows that ϕ_n is differentiable and that it is

$$\frac{d}{dt}\phi_n(t) = -\phi_{n+1}(t).$$

Thus $\overline{F}^{(1)}(t)$ admits derivatives of any order and

$$\frac{d^n}{dt^n}\overline{F}^{(1)}(t) = (-1)^n \phi_n(t); \tag{1.45}$$

furthermore it is obviously such that

$$\overline{F}^{(1)}(0) = 1. \tag{1.46}$$

A function satisfying the conditions (1.45) is *completely monotone.* Any completely monotone function satisfying the additional condition (1.46) is the Laplace transform of a probability distribution Π_0 on $[0, +\infty]$. i.e., for some Π_0 we can write

$$\overline{F}^{(1)}(t) = \int_0^\infty \exp\{-\theta t\}d\Pi_0(\theta), \tag{1.47}$$

see e.g. Feller (1971), pg. 439. The equation (1.43) is then obtained by combining Equation (1.44) with (1.45) and (1.47).   □

A more general proof can be found in Diaconis and Freedman (1987).

Remark 1.45. (Mathematical meaning of Theorem 1.44). We say that a joint density function is *Schur-constant* if it satisfies condition 1.42. Joint (predictive) densities of n conditionally independent, exponentially distributed nonnegative random variables (i.e. densities satisfying Equation (1.43)) are special cases of Schur-constant densities. Although these are not the only cases, they however constitute the only cases compatible with the condition of being marginal of N-dimensional Schur-constant densities for any N.

Another way to state the same fact is that denumerable products of exponential distributions are the extreme points of the convex set of all the distributions on $\mathbb{R}^\infty_+$ with absolutely continuous finite dimensional marginals satisfying the condition (1.42).

Remark 1.46. (Statistical meaning of Theorem 1.44). As we shall see in Chapter 4, the property (1.42) is equivalent to a very significant condition of invariance in the context of survival analysis, which will be interpreted as a "no-aging property" for a n-tuple of lifetimes. Theorem 1.44 states that, in the case of infinite extendibility, imposing such a condition for any n is equivalent to assessing the "lack of memory" model (1.43).

1.4.1   Parametric models and prediction sufficiency

As mentioned above, certain parametric models for sequences of exchange-able random variables can be justified in terms of de Finetti-type theorems. Generally, de Finetti-type theorems can be approached from a number of different mathematical viewpoints (see e.g. Dynkin, 1978; Dawid,1982; Lauritzen, 1984; Diaconis and Freedman, 1987; Ressel, 1985; Lauritzen, 1988 and references cited therein). Several examples are presented in Diaconis and Freedman (1987). A further, although not exhaustive, bibliography is given at the end of this Chapter.

In what follows we shall not analyze specific parametric models. Rather we aim to give just a mention of the possible role of the notion of *prediction sufficiency* within the framework of de Finetti-type results.

Let $X_1, X_2, ...,$ be an N-extendible family of exchangeable quantities (possibly $N = \infty$); we limit ourselves to consider the case when $(X_1, X_2, ..., X_m)$ admit a probability density $f^{(m)}$, for $m = 1, 2, ..., N$. Let Z_m be a given statistic of $X_1, X_2, ..., X_m$.

Definition 1.47. Z_m is a *prediction sufficient statistic* if, for any n such that $m + n \leq N$,

$$(X_{m+1}, ..., X_{m+n}) \text{ and } (X_1, ..., X_m)$$

are conditionally independent given $Z_m(X_1, ..., X_m)$.

The existence of statistics can be seen as a condition of invariance, stronger than exchangeability, for the joint densities and we have the following:

Proposition 1.48. *The above definition is equivalent to the existence of functions $\psi_{m,n}$ such that, for $\mathbf{x}' \in R^m$, and $\mathbf{x} \equiv (x_{m+1}, ..., x_{m+n}) \in \mathbb{R}^n$, one has*

$$f^{(m+n)}(\mathbf{x}', \mathbf{x}) = \psi_{m,n}(Z_m(\mathbf{x}'); \mathbf{x}) f^{(m)}(\mathbf{x}') \tag{1.48}$$

Proof. Conditional independence among $(X_{m+1}, ..., X_{m+n})$ and $(X_1, ..., X_m)$, given $Z_m(X_1, ..., X_m)$ means that the conditional density $f^{(n)}(\mathbf{x}|X_1 = x'_1, ..., X_m = x'_m)$ depends on $\mathbf{x}'$ only through the value $Z_m(\mathbf{x}')$ and then

$$f^{(n)}(\mathbf{x}|X_1 = x'_1, ..., X_m = x'_m) = \psi_{m,n}(\mathbf{x}, Z_m(\mathbf{x}')) .$$

Whence Equation (1.48) follows for the joint density

$$f^{(m+n)}(\mathbf{x}', \mathbf{x}) = f^{(m)}(\mathbf{x}') f^{(n)}(\mathbf{x}|X_1 = x'_1, ..., X_m = x'_m)$$

of $X_1, ..., X_m, X_{m+1}, ..., X_{m+n}$. □

Remark 1.49. Definition 1.47 does not require infinite extendibility. Even in the case of infinite extendibility, anyway, the concept of prediction sufficiency is formulated without referring to a "parameter", i.e. to a random variable Θ, with respect to which $X_1, X_2, \ldots$ are conditionally i.i.d.

This feature of the definition is indeed convenient in the present setting, where the problem of interest is just proving the existence of a parameter, in terms of conditions of invariance or of sufficiency.

For our purposes, we are interested in the study of infinitely extendible families admitting chains of prediction sufficient statistics

$$Z_1(X_1), Z_2(X_1, X_2), \ldots, Z_m(X_1, X_2, \ldots, X_m), \ldots.$$

More specifically our attention is to be focused on sequences of prediction sufficient statistics with the following two additional properties:

i) for any m, $Z_m(x_1, \ldots, x_m)$ is *symmetric*, i.e.

$$Z_m(x_1, \ldots, x_m) = Z_m(x_{j_1}, \ldots, x_{j_m}) \tag{1.49}$$

for any permutation $(j_1, \ldots, j_m)$ of $(1, \ldots, m)$

ii) the sequence $\{Z_m\}_{m=1,2,\ldots,N}$ is *algebraically transitive* i.e. mappings ψ_m exist such that

$$Z_{m+1}(x_1, \ldots, x_m, x_{m+1}) = \psi_{m+1}(Z_m(x_1, \ldots, x_m), x_{m+1})$$

i.e., such that for all x, the following implication holds

$$Z_m(x_1', \ldots, x_m') = Z_m(x_1'', \ldots, x_m'') \Rightarrow$$

$$Z_{m+1}(x_1', \ldots, x_m', x) = Z_{m+1}(x_1'', \ldots, x_m'', x). \tag{1.50}$$

Before continuing, it is convenient to reconsider known cases in this framework.

Example 1.50. Let $X_1, X_2, \ldots, X_N$ be $\{0, 1\}$-valued exchangeable quantities and consider, for $m = 1, 2, \ldots, N - 1$, the statistic

$$S_m(x_1, \ldots, x_m) = \sum_{i=1}^{m} x_i.$$

S_m obviously is a prediction sufficient statistic, as is shown by the formula (1.17). S_m is trivially a symmetric function and the property (1.50) is satisfied in that

$$S_m(x_1, \ldots, x_m, x_{m+1}) = S_m(x_1, \ldots, x_m) + x_{m+1}.$$

Example 1.51. Consider a parametric model in which $f^{(n)}$ $(n \geq 1)$ is of the form (1.38) with $\{g(x|\theta)\}_{\theta \in L}$ in particular being a regular exponential family of order 1. Namely we assume that, for suitable real functions a, b, H, K,

$$g(x|\theta) = a(x)b(\theta)\exp\{H(x)K(\theta)\} \tag{1.51}$$

This means that, for $n = 1, 2, ...$,

$$f^{(n)}(x_1, ..., x_n) =$$

$$\prod_{i=1}^{n} a(x_i) \int_L [b(\theta)]^n \cdot \exp\{K(\theta) \cdot Z_n(x_1, ..., x_n)\} d\Pi_0(\theta) \tag{1.52}$$

where $Z_n : \mathbb{R}^n \to \mathbb{R}$ is defined by $Z_n(x_1, ..., x_n) \equiv \sum_{i=1}^{m} H(x_i)$. $Z_m(X_1, ..., X_m)$ is sufficient in the usual sense of Bayesian statistics (see e.g. Zacks, 1971). One can easily show that, for any m, Z_m is also prediction sufficient. It is immediate that Z_m satisfies the conditions (1.49) and (1.50).

Example 1.52. Let $X_1, ..., X_N$ be non-negative exchangeable quantities with a joint density function satisfying condition (1.42). Letting, for $m = 1, ..., N-1$,

$$S_m(x_1, ..., x_m) \equiv \sum_{i=1}^{m} x_i$$

it is immediately seen that S_m is a prediction sufficient statistic, satisfying the conditions (1.49) and (1.50).

Example 1.53. Consider now an arbitrary family $X_1, ..., X_N$ of (real-valued) exchangeable quantities and let

$$Z_m(x_1, ..., x_m) \equiv \Phi_m(x_1, ..., x_m)$$

where $\Phi_m(x_1, ..., x_m)$ denotes the vector $(x_{(1)}, ..., x_{(m)})$ of order the statistics of $(x_1, ..., x_m)$.

Due to exchangeability, it is immediate that $Z_m(X_1, ..., X_m)$ is a prediction sufficient statistic. The conditions (1.49) and (1.50) are trivially satisfied.

Remark 1.54. Example 1.53 shows that we can always find at least one sequence of prediction sufficient statistics satisfying the additional properties (1.49) and (1.50). This is the sequence of order statistics. However this may not be the only one, as Examples 1.50, 1.51 and 1.52 show.

Let us now come back to the discussion, presented in the previous section, concerning the classical de Finetti representation result 1.32. We can see that the sequence of order statistics plays there a role similar to that of the specific sequence of statistics

$$S_n(X_1, ..., X_n) = \sum_{i=1}^{n} X_i$$

for the case of exchangeable events. In this respect, recall also the Remark 1.37.

Starting from these considerations, we can in general remark that any sequence of prediction sufficient statistics can play a similar role, whenever it satisfies the conditions (1.49) and (1.50).

Specifically, one can prove in fact that the sequence $\{Z_m\}_{m=1,2,...}$ allows us to represent $X_1, X_2, ...$ as a sequence of conditionally i.i.d. random quantities: for fixed n, $X_1, X_2, ..., X_n$ tend to become conditionally i.i.d. given $Z_N(X_1, X_2, ..., X_N)$, when $N \to \infty$.

In this respect, the following fact is relevant: consider the variables

$$Z_1(X_1), ..., Z_{N-1}(X_1, ..., X_{N-1})$$

for exchangeable quantities $X_1, ..., X_N$.

By taking into account the conditional independence of $(X_1, ..., X_m)$ and $(X_{m+1}, ..., X_N)$ given $Z_m(X_1, ..., X_m)$, one can show that $Z_1(X_1), Z_2(X_1, X_2), ...$ have properties which generalize those specified by equations (1.8) and (1.9).

More precisely, it is only a technical matter to show the following result.

Proposition 1.55. *For m, n such that $m + n < N$, for arbitrary (measurable) sets B_m and B_{m+n}, the following properties hold:*

$$P\{Z_{m+n}(X_1, ..., X_{m+n}) \in B_{m+n} | X_1, ..., X_m\} =$$

$$P\{Z_{m+n}(X_1, ..., X_{m+n}) \in B_{m+n} | Z_1, ..., Z_m\} =$$

$$P\{Z_{m+n}(X_1, ..., X_{m+n}) \in B_{m+n} | Z_m(X_1, ..., X_m)\} \tag{1.53}$$

and

$$P\{Z_m(X_1, ..., X_m) \in B_m | Z_{m+n}, X_{m+n+1}, ..., X_N\} =$$

$$P\{Z_m(X_1, ..., X_m) \in B_m | Z_{m+n}(X_1, ..., X_{m+n})\} \tag{1.54}$$

When, in the infinitely extendible case, the sequence

$$Z_1(X_1), ..., Z_N(X_1, ..., X_N), ...$$

converges to a random variable Θ, one can find technical conditions under which $X_1, X_2, ...$ are conditionally i.i.d. with respect to Θ.

Remark 1.56. Results of the type mentioned above allow us to characterize, and justify, specific finite-dimensional parametric models of the type (1.35), by assigning the sequence of sufficient statistics.

Note also that, in such cases, the "parameter" Θ, being the limit of the sequence $\{Z_N(X_1, ..., X_N)\}$, is given a meaning as a function of the whole sequence of the observable quantities $X_1, X_2, ...$ (see also the discussion in Dawid, 1986).

It is of interest also to point out that, in those cases, it can happen that, for any n, $(X_1, ..., X_n)$ and Θ are conditionally independent given $Z_n(X_1, ..., X_n)$, i.e. $Z_n(X_1, ..., X_n)$ is sufficient for Θ, according to the usual definition.

1.5   Exercises

Exercise 1.57. Prove Proposition 1.3.

Exercise 1.58. For n exchangeable events $E_1, ..., E_n$, let $S_n = \sum_{i=1}^{n} X_i$ be the number of successes. Check that $E_1, ..., E_n$ are independent if and only if the distribution of S_n is binomial $b(n, \theta)$, for some θ.

Exercise 1.59. Let $f(x_1, x_2)$ be the symmetric joint density function defined by

$$f(x_1, x_2) = \begin{cases} \alpha & for \ (x_1, x_2) \in C \cap A \\ \gamma & for \ (x_1, x_2) \in C \cap \overline{A} \\ 0 & for \ (x_1, x_2) \in \overline{C} \end{cases}$$

where $\alpha, \gamma > 0$,

$$A \equiv \{(x_1, x_2) | x_1 \cdot x_2 \geq 0\}$$

$$C \equiv \{(x_1, x_2) | -1 \leq x_1 + x_2 \leq 1, -z \leq x_1 - x_2 \leq z\}, \ z > 2.$$

(see Figure 2)

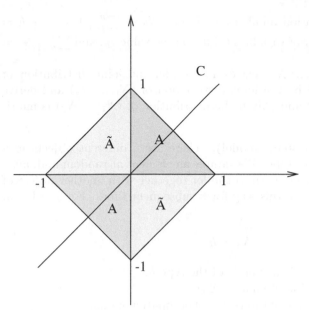

Figure 2

Show that f is not 3-extendible.

Hint: note that a necessary condition for 3-extendibility is the possibility of finding out a domain $\Omega_3 \subset \mathcal{R}^3$, such that

(a) for any permutation (i_1, i_2, i_3) of $(1, 2, 3)$,

$$(x_1, x_2, x_3) \in \Omega_3 \Rightarrow (x_{i_1}, x_{i_2}, x_{i_3}) \in \Omega_3$$

(b)

$$\Omega_2 = C$$

where

$$\Omega_2 \equiv \{(x_1, x_2) \in \mathbb{R}^2 \mid \ \xi \in \mathbb{R} \text{ exists such that } (x_1, x_2, \xi) \in \Omega_3\}$$

Exercise 1.60. (Continuation) For the above density, prove that we can have cases of positive correlation or negative correlation, by suitably choosing the constants a and γ.

The next three exercises concern discrete random variables. Consider random variables $X_1, ..., X_n$ and let $\mathcal{X} \equiv \{z_1, ..., z_d\}$ be the set of their possible values. Note that the condition of exchangeability in this cases becomes

$$P\{X_1 = x_1, ..., X_n = x_n\} = P\{X_1 = x_{j_1}, ..., X_n = x_{j_n}\}$$

where $x_1, ..., x_n \in \mathcal{X}$ and $(x_{j_1}, ..., x_{j_n})$ is any permutation of $(x_1, ..., x_n)$.

Exercise 1.61. ("Occupation numbers"). Define $N_h \equiv \sum_{j=1}^{n} \mathbf{1}_{\{X_j = z_h\}}, h = 1, 2, ..., d$; N_h is the number of variables taking on the value z_h, and $\sum_{h=1}^{d} N_h = n$.

Prove that, when $X_1, ..., X_n$ are exchangeable, the joint distribution of $(X_1, ..., X_n)$ is determined by the joint distribution of $(N_1, ..., N_d)$ and derive that $X_1, ..., X_n$ are i.i.d. if and only if the distribution of $(N_1, ..., N_d)$ is multinomial.

Exercise 1.62. ("Bose-Einstein" model). There are d different objects in a lot; any object is of different type. We sample an element at random and, after inspection, we reinsert the object into the lot together with another object of the same type; we continue in this way for n subsequent times. For $j = 1, ...n$, $h = 1, 2, ..., d$, define

$$X_j = h$$

if the object sampled at the j-th time is of the type h.

Find the joint distribution of $(X_1, ..., X_n)$.

(Hint: find the distribution of the vector of occupation numbers $(N_1, ..., N_d)$, defined in the previous exercise.)

Exercise 1.63. Let $X_1, ..., X_n$ be exchangeable random variables taking values in $\mathbb{N} \equiv \{1, 2, ...\}$ and such that

$$P\{\mathbf{X} = \mathbf{x}\} = \psi_n \left(\sum_{j=1}^{n} x_j \right).$$

Compute the distribution of $S_n \equiv \sum_{j=1}^{n} X_j$ and the conditional distribution of $(X_1, ..., X_n)$ given $\{S_n = s\}$. Check that this does not depend on the particular choice of ψ_n.

The next exercise concerns continuous variables with "Schur-constant" densities, i.e. satisfying the invariance condition

$$f^{(n)}(\mathbf{x}) = \phi_n \left(\sum_{j=1}^{n} x_j \right).$$

Such a condition can be seen as the most direct analog, for the case of continuous random variables, of the notion of exchangeability for binary random variables. Again we denote $S_n \equiv \sum_{j=1}^{n} X_j$.

Exercise 1.64. a) Prove that S_n has a density of the form

$$f_{S_n}(s) = K_n(s)\phi_n(s), \quad s \geq 0$$

where the function K_n does not depend on ϕ_n.

b) Prove that $X_1, ..., X_n$ are i.i.d., exponentially distributed, if and only the distribution of S_n is gamma with parameters n and θ for some $\theta > 0$:

$$f_{S_n}(s) \propto s^{n-1} \exp\{-\theta s\}, \quad s \geq 0$$

and derive that

$$K_n(s) = \frac{s^{n-1}}{(n-1)!}.$$

c) Prove that, for $1 \leq m < n$, the conditional density of $(X_1, ..., X_m)$ given $\frac{S_n}{n}$ is the same as in the special case when $X_1, ..., X_n$ are i.i.d., exponentially distributed.

d) In terms of the function ϕ_n, write the conditional density of X_n, given $\{X_1 = x_1, ..., X_{n-1} = x_{n-1}\}$.

Exercise 1.65. Let $(Y_1, ..., Y_n)$ a vector of non-negative random variables with a (non-exchangeable) joint density f_Y. Prove that we can find an exchangeable density $f^{(n)}$ for random variables $(X_1, ..., X_n)$ such that the vectors of order statistics of $(Y_1, ..., Y_n)$ and $(X_1, ..., X_n)$ have the same joint probability distribution.

Exercise 1.66. Prove Proposition 1.48.

Exercise 1.67. Consider a parametric model with $\{g(x|\theta)\}_{\theta \in L}$ being a regular exponential family of order 1 as in the Example 1.51. Check that $Z_n(x_1, ..., x_n) \equiv \sum_{i=1}^{m} H(x_i)$ is prediction sufficient.

1.6   Bibliography

Aldous, D.J. (1983). *Exchangeability and related topics.* Springer Series Lecture Notes in Mathematics, 1117.

Bernardo, J. and Smith, A. F. M. (1994). *Bayesian theory.* John Wiley & Sons, New York.

Bruno, G. and Gilio, A. (1980). Application of the simplex method to the fundamental theorem for the probabilities in the subjective theory. (Italian). *Statistica*, 40, 337–344

Cassel, D., Särndal, C.E. and Wretman, J.H. (1977). *Foundations of inference in survey sampling.* John Wiley & Sons, New York.

Chow, Y. S. and Teicher H. (1978). *Probability Theory. Independence, Interchangeability, Martingales.* Springer Verlag, New York.

Cifarelli, M.D. and Regazzini, E. (1982), Some considerations about mathematical statistics teaching methodology suggested by the concept of exchangeability. In *Exchangeability in Probability and Statistics*, G. Koch and F. Spizzichino, Eds. North-Holland, Amsterdam.

Dawid, A. P. (1979). Conditional independence (with discussion). *J. Roy. Statist. Soc.* Ser. B 41 (1979), no. 1, 1–31.

Dawid, A.P. (1982). Intersubjective statistical models. In *Exchangeability in Probability and Statistics*, G. Koch and F. Spizzichino, Eds. North-Holland, Amsterdam.

Dawid, A.P. (1985). Probability, symmetry and frequency. *British J. Philos. Sci.*, 36, 107- 112.

Dawid, A.P. (1986). A Bayesian view of statistical modelling. In *Bayesian inference and decision techniques*, P.K. Goel and A. Zellner, Eds. Elsevier, Amsterdam.

de Finetti, B. (1931). Funzione caratteristica di un fenomeno aleatorio. *Atti della R. Accademia Nazionale dei Lincei*, Ser. 6, Mem. Cl. Sci. Mat. Fis. Nat., 4, 251-299.

de Finetti, B. (1937). La Prevision: ses lois logiques, ses sources subjectives. *Ann. Inst. Henry Poincare*,7, 1-68.

de Finetti, B. (1952). Gli eventi equivalenti ed il caso degenere. *G. Ist. Ital. Attuari*, Vol. 15, 40-64.

de Finetti, B. (1964). Alcune osservazioni in tema di "suddivisione casuale". *G. Ist. Ital. Attuari*, 27, 151–173.

de Finetti, B. (1970). *Teoria delle Probabilita'*. Einaudi, Torino. English translation: *Theory of Probability*, John Wiley and Sons, New York (1974).

de Finetti, B. (1972). *Probability, induction and statistics. The art of guessing.* Wiley Series in Probability and Mathematical Statistics. John Wiley & Sons, London-New York-Sydney.

Diaconis, P. and Freedman, D. (1987). A dozen de Finetti-style results in search of a theory. *Ann. Inst. Henry Poincare* 23, 397-423.

Diaconis, P. and Freedman, D. (1990). Cauchy's equation and de Finetti's theorem. *Scand. J. Stat.* 17, 235-250.

Diaconis, D. and Ylvisaker, D. (1985). Quantifying prior probabilities. In *Bayesian Statistics*, Vol. 2, Bernardo, De Groot, Lindley, Smith Eds.

Dubins, L. and Freedman, D. (1979). Exchangeable process need not be mixtures of identically distributed random variables. *Z. Wahrsch. verw. Gebiete*, 33, 115-132.

Dynkin, E.B. (1978). Sufficient statistics and extreme points. *Ann. Probab.*, 6, 705-730.

Ericson, W.A. (1969) Subjective Bayesian models in sampling finite populations. *J. Roy. Statist. Soc Ser. B.* 31, 195-233.

Feller, W. (1968) *An Introduction to probability theory and its applications.* Vol. 1. Third edition John Wiley & Sons, New York-London-Sydney.

Feller, W. (1971). *An Introduction to probability theory and its applications.* Vol. 2. John Wiley & Sons, New York-London-Sydney.

Flourens, J.P., Mouchart, M.J. and Rolin, M. (1990). *Elements of Bayesian statistics.* Marcel Dekker, New York-Basel.

Heath, D. and Sudderth, W. (1976). De Finetti's theorem for exchangeable random variables. *Amer. Statist.* 30, no. 4, 188–189.

Hewitt, E. and Savage, L.J. (1955). Symmetric measures on cartesian products. *Trans. Am. Math. Soc.*, 80, 470-501.

Hill, B.M. (1993). Dutch books, the Jeffrey-Savage theory of hypothesis testing and Bayesian reliability. In *Reliability and Decision Making*, R.E. Barlow, C.A. Clarotti, F. Spizzichino Eds., Chapman and Hall, London, 31-85.

Jackson, M., Kalai, E. and Smorodinsky, R. (1999). Bayesian representation of stochastic processes under learning: de Finetti revisited. *Econometrica* 67 , no. 4, 875–893.

Kendall, D.G. (1967). On finite and infinite sequences of exchangeable events. *Studia Scient. Math. Hung.*, 2, 319-327.

Kyburn, H. and Smokler, H. (Eds.) (1964). *Studies in subjective probability.* John Wiley & Sons, New York.

Koch, G. and Spizzichino, F. (Eds.) (1982). *Exchangeability in probability and statistics.* North-Holland, Amsterdam.

Lad, F. (1996). *Operational subjective statistical methods. A mathematical, philosophical, and historical introduction.* John Wiley & Sons, New York.

Lauritzen, S. L. (1974). Sufficiency, prediction and extreme models. *Scand. J. Stat.*, 1, 128-134.

Lauritzen, S. L. (1984). Extremal point methods in statistics. *Scand. J. Stat.*, 11, 65-91.

Lauritzen, S. L. (1988). *Extremal families and systems of sufficient statistics.* Springer Verlag, New York.

LeCam, L. (1958). Les propriétés asymptotiques de solution de Bayes. *Publications de l'Institut de Statistique de l'Université de Paris,* 7, 1419-1455.

Lindley, D.V. and Novick, M.R. (1981). The role of exchangeability in inference. *Ann. Stat.* 9, 45-58.

Loève, M. (1963). *Probability theory* Third edition Van Nostrand, Princeton.

Olshen, R. (1974). A note on exchangeable sequences. *Z. Wahrsch. verw. Gebiete,* 28, 317-321.

Picci, G. (1977). Some connections between the theory of sufficient statistics and the identification problem. *SIAM J. Appl. Math.*, 33, 383-398.

Piccinato, L. (1993). The Likelihood Principle in Reliability Analysis. In *Reliability and Decision Making*, R. E. Barlow, C. A. Clarotti and F. Spizzichino, Eds, Chapman & Hall, London.

Piccinato, L. (1997). *Metodi per le decisioni statistiche.* Springer Italia, Milano.

Regazzini, E. (1988). Subjective Probability. In *Encyclopedia of Statistical Sciences*, Vol. 9, N. Johnson and S. Kotz Eds., John Wiley & Sons, New York.

Renyi, A. and Revesz, P. (1963). A study of sequences of equivalent events as special stable sequences. *Publ Math. Debrecen,* 10, 319-325.

Ressel, P. (1985). de Finetti-type theorems: an analytical approach. *Ann. Probab.*, 13, 898-922.

Ryll-Nardewski (1955). On stationary sequences of random variables and de Finetti's equivalence. *Coll. Math.*, 4, 149-156.

Savage, L. J. (1972) *The foundations of statistics*, 2nd revised edition. Dover, New York.

Scozzafava, R. (1982). Exchangeable events and countable disintegrations. In *Exchangeability in probability and statistics*, G. Koch and F. Spizzichino, Eds. North-Holland, Amsterdam, 297-301.

Spizzichino, F. (1978). Statistiche sufficienti per famiglie di variabili aleatorie scambiabili. In *Atti Convegno su Induzione, Probabilita' e Statistica*, Universita' di Venezia.

Spizzichino, F. (1982). Extendibility of symmetric probability measures. In *Exchangeability in probability and statistics*, G. Koch and F. Spizzichino, Eds. North-Holland, Amsterdam, 313-320.

Spizzichino, F. (1988). Symmetry conditions on opinion assessment leading to time-transformed exponential models. In *Prove di durata accelerate e opinioni degli esperti in Affidabilita'*, C.A. Clarotti and D. Lindley, Eds. North Holland.

Sugden, R. (1979). Inference on symmetric functions of exchangeable populations. *J. R. Stat. Soc.* B, 41, 269-273.

Urbanik, K. (1975). Extreme points methods in probability. In *Proceedings of the Winter School on Probability, Karpacz, 1975*.

Zacks, K. (1971). *The theory of statistical inference.* John Wiley & Sons, New York.

Chapter 2

Exchangeable lifetimes

2.1 Introduction

In this chapter we shall be dealing with basic aspects of the distribution of n non-negative exchangeable random quantities that will be denoted by $T_1, ..., T_n$. $T_1, ..., T_n$ can be interpreted as lifetimes either of elements in a biological population or of units of an industrial production; elements or units will be sometimes denoted by $U_1, ..., U_n$ and commonly referred to as *individuals*.

The assumption that $T_1, ..., T_n$ are exchangeable embodies the idea that $U_1, ..., U_n$ are similar, at least as far as our state of information is concerned: we expect that $U_1, ..., U_n$ will have different performances and that their lifetimes $T_1, ..., T_n$ will take different values, but we have no reason to assess for them different probability laws.

A list of different, simple cases giving rise to exchangeable lifetimes follows; a more precise description of corresponding probabilistic models will be given in the next sections. These models will also provide the basis for most of the examples and discussions to be presented from now on.

Example 2.1. (Conditionally i.i.d. lifetimes). This is the most common case. $U_1, ..., U_n$ are similar and there is no physical interaction among them; however, their lifetimes are influenced by the values taken by one or more physical quantities, such as intrinsic strength of material or temperature, pressure and, more in general, environmental conditions (similar cases can easily be conceived in the biomedical field). The vector formed by the latter quantities is denoted by the symbol Θ. Suppose that the actual value of Θ is constant during the lives of $U_1, ..., U_n$, but it is unobservable to us and there is an uncertainty to it. Then we assess a probability distribution for Θ. $T_1, ..., T_n$ may be assumed to be conditionally independent, identically distributed given Θ, and thus exchangeable, but they are not independent. During the observation of progressive survivals and/or of failures of $U_1, ..., U_n$ a continuing process of learning about Θ takes place; this means that we continuously update the distribution of Θ.

Example 2.2. (Change-point models). Θ is a non-negative quantity, measuring the level of the instantaneous stress on the individuals $U_1, ..., U_n$; the operative lives of $U_1, ..., U_n$ start simultaneously at time $t = 0$ and the value taken by Θ, at $t = 0$, is known to be equal to θ_1.

At a time $\sigma > 0$, Θ changes its value to θ_2. We do not observe σ directly; however, we observe progressive failures of $U_1, ..., U_n$. $T_1, ..., T_n$ are then assumed to be conditionally independent, identically distributed given σ. What we have observed up to any time $t > 0$ will be used to update our judgment about the event $\{\sigma \leq t\}$ and about the residual lifetimes $T_i - t$.

This is a special case of what is known as the "disruption" or "change point" problem.

Furthermore we can combine the case above with the situation of Example 2.1, by regarding θ_1 and θ_2 as unknown to us.

More generally we might think of cases where the value of Θ evolves in time, giving rise to a (non-observable) stochastic process Θ_t.

This situation may become much more general and complicated than that in the previous example; however, $T_1, ..., T_n$ remain exchangeable as far as, for all $t > 0$, all individuals surviving at t have a same probabilistic behavior, conditional on the process Θ_t.

Example 2.3. (Heterogeneous populations). Among the n individuals $U_1, ..., U_n$, some are *strong* and some are *weak*. We neither know the identity of weak individuals nor the total number of them and, for $j = 1, ..., n$, we introduce the binary random variables defined as follows:

$$K_j = \begin{cases} 1 & U_j \text{ is "weak"} \\ 0 & U_j \text{ is "strong"} \end{cases}$$

We think of the situation where, if we knew the condition $\{K_j = i\}$ $(i = 1, 2)$, of the individual U_j, then its lifetime T_j would be independent of the other lifetimes and distributed according to a given distribution $G_i(t)$.

Thus $T_1, ..., T_n$ are independent, though non-identically distributed, given $K_1, ..., K_n$. Under the condition that $K_1, ..., K_n$ are exchangeable, $T_1, ..., T_n$ are also exchangeable, but not infinitely extendible, in general, according to the definition given in 1.28.

$K_1, ..., K_n$ are i.i.d. if and only if the total number M of weak individuals has a binomial distribution $b(n, p)$ for some p (see Exercise 1.58); in such a case $T_1, ..., T_n$ are i.i.d. as well.

The extension of this example to the case when there are more than two categories is rather straightforward; we can in fact consider the case of exchangeable random variables $K_1, ..., K_n$ which are not necessarily binary (see Subsection 4.3 in Chapter 4).

Example 2.4. (Finite exchangeability). The individuals $U_1, ..., U_n$ are similar and behave independently of each other. At a certain step, we get the information $\{S(T_1, ..., T_n) = s\}$, where $S(\mathbf{t}) = \sum_{i=1}^{n} t_i$ or S is any other symmetric statistic.

As noticed in Remark 1.43, the distribution of $(T_1, ..., T_n)$ conditional on this event is exchangeable; however, $T_1, ..., T_n$ are not independent. Furthermore their (n-dimensional) joint distribution does not admit a density function.

This is a case of finite-extendibility.

Example 2.5. (Time-transformed exponential models). The lifetime of each of the individuals $U_1, ..., U_n$ in a population is a deterministic function of a certain resource A available to it, so that we can put $T_i = \rho(X_i)$, where X_i denotes the initial amount of the resource A initially owned by U_i. The total initial amount of A in the population is a random quantity $X = \sum_{i=1}^{n} X_i$.

We assume that, conditional on the hypothesis $\{X = x\}$, the joint distribution of $X_1, ..., X_n$ is uniform over the simplex $\sum_{i=1}^{n} X_i = x$; in other words $X_1, ..., X_n$ are the *spacings* of a random partition of the interval $[0, X]$.

$T_1, ..., T_n$ are exchangeable; their joint distribution is infinitely extendible or not depending on the marginal distribution of X.

Example 2.6. (Exchangeable load-sharing models). The individuals are units which are similar and are installed to work simultaneously in the same system or to supply the same service (think of the set of copy-machines in a university department). Then they share a similar load and, when one of them fails, the others undergo an increased stress. This is then a case of positive dependence. The notion of multivariate conditional hazard rate, which will be defined later on, provides a convenient tool to describe mathematically such models.

Example 2.7. (Exchangeable common mode failure models). Think again of units which are similar and work simultaneously.

Now we consider the case when they work independently of each other. However they are sensible to the same shock. The shock arrives at a random time; at any instant t, the hazard rate of the arrival of the shock is independent of the number and identity of the units failed up to t. When the shock arrives all the surviving units fail simultaneously. This is again a case of positive dependence.

Also in this case the joint distribution does not admit a density.

From now on in this chapter we recall the terminology and fundamental probabilistic concepts which are specific to the treatment of non-negative random quantities. In the remainder of the present section we consider the one-dimensional case, while the next three sections will be devoted to the case of several variables. In particular, the concept of multivariate conditional hazard rate will be introduced in Section 2.3 and specific aspects of it will be analyzed in Section 2.4.

In the one-dimensional cases, fundamental concepts are those of the *conditional survival function of a residual life-time at an age* s and, for distributions admitting a density function, of *hazard rate function.*

For a distribution function F, such that $F(0) = 0$, let U be an individual whose lifetime T has the survival function

$$\overline{F}(s) \equiv P\{T > s\} = 1 - F(s).$$

The conditional survival function of the residual lifetime of U at age s is

$$\overline{F}_s(\tau) \equiv P\{T - s > \tau | T > s\} = \frac{\overline{F}(\tau + s)}{\overline{F}(\tau)}, \quad s > 0, \tau > 0 \qquad (2.1)$$

In the absolutely continuous case, when $F(s) = \int_0^s f(t)dt \; (s > 0)$, the hazard rate function is defined by

$$r(s) = \lim_{\delta \to 0^+} \frac{P\{T - s \leq \delta | T > s\}}{\delta} = \lim_{\delta \to 0^+} \frac{1 - \overline{F}_s(\delta)}{\delta} =$$

$$\frac{f(s)}{\overline{F}(s)} = -\frac{d}{ds} \log \overline{F}(s) \qquad (2.2)$$

By introducing the *cumulative hazard function* $R(s) \equiv \int_0^s r(\xi)d\xi$, one can write

$$\overline{F}(s) = \exp\{-R(s)\}$$

$$f(t) = r(t) \exp\{-R(t)\} \qquad (2.3)$$

$$\overline{F}_s(\tau) = \exp\{R(s) - R(s + \tau)\}$$

Example 2.8. The exponential distribution with parameter θ ($\overline{F}(s) = \exp\{-\theta s\}$) is characterized by the properties

$$\overline{F}_s(\tau) = \exp\{-\theta\tau\} = \overline{F}(\tau), r(s) = \theta, \; \forall s > 0, \forall \tau > 0.$$

Example 2.9. As in Example 2.1, consider the case when T has a conditional survival function $\overline{G}(s|\theta)$ depending on the value θ of some physical quantity Θ, where Θ is a random quantity taking values in a set L and with an initial density π_0. The "predictive" survival and density function of T are then, respectively,

$$\overline{F}(s) = \int_L \overline{G}(s|\theta)\pi_0(\theta)d\theta; \quad f(t) = \int_L g(t|\theta)\pi_0(\theta)d\theta$$

so that, for $s > 0, \tau > 0$,

$$\overline{F}_s(\tau) = \frac{\overline{F}(\tau + s)}{\overline{F}(s)} = \frac{\int_L \overline{G}(\tau + s|\theta)\pi_0(\theta)d\theta}{\int_L \overline{G}(s|\theta)\pi_0(\theta)d\theta} =$$

$$\frac{\int_L \frac{\overline{G}(\tau+s|\theta)}{\overline{G}(s|\theta)}\overline{G}(s|\theta)\pi_0(\theta)d\theta}{\int_L \overline{G}(s|\theta)\pi_0(\theta)d\theta} = \int_L \overline{G}_s(\tau|\theta)\pi_s(\theta|s)d\theta \qquad (2.4)$$

where $\pi_s(\theta|s)$ is defined by

$$\pi_s(\theta|s) = \frac{\overline{G}(s|\theta)\,\pi_0(\theta)}{\int_L \overline{G}(s|\theta)\pi_0(\theta)d\theta}.$$

By Bayes' formula, $\pi_s(\theta|s)$ can be interpreted as the conditional density of Θ, given the observation of the survival $(T > s)$.

Denoting by $r(s|\theta)$, $R(s|\theta)$ the hazard rate function and the cumulative hazard function of $\overline{G}(s|\theta)$, respectively, we can also write

$$\overline{F}(s) = \int_L \exp\{-R(s|\theta)\}\pi_0(\theta)d\theta$$

$$f(t) = \int_L r(t|\theta)\exp\{-R(t|\theta)\}\pi_0(\theta)d\theta$$

and the predictive (unconditional) hazard rate function is

$$r(t) = \frac{f(t)}{\overline{F}(t)} = \frac{\int_L r(t|\theta)\exp\{-R(t|\theta)\}\pi_0(\theta)d\theta}{\int_L \exp\{-R(t|\theta)\}\pi_0(\theta)d\theta} = \int_L r(t|\theta)\pi_t(\theta|t)d\theta. \qquad (2.5)$$

The predictive joint survival function and the predictive density, respectively, are given by mixtures of the conditional survival functions and densities, π_0 being the compounding density. Notice that also for the predictive hazard rate $r(s)$ we found that it is a mixture of the conditional hazard rate functions $r(t|\theta)$. But, differently from what happens for $\overline{F}$ and f, the compounding density this time is $\pi_t(\theta|t)$, which varies with t. This fact will have important consequences in the study of "aging properties" of mixtures.

Example 2.10. Think of a lifetime T with a (predictive) hazard rate function $r(t)$; suppose that, at a certain step, we get the information $\{T < \tau\}$ (for a given $\tau > 0$). The conditional survival function is, for $s < \tau$,

$$\overline{F}(s|T < \tau) = \frac{\overline{F}(s) - \overline{F}(\tau)}{1 - \overline{F}(\tau)}$$

and then the conditional hazard rate function becomes, for $t < \tau$,

$$r(t|T < \tau) = r(t) \frac{1}{1 - \frac{F(\tau)}{F(t)}} \tag{2.6}$$

Example 2.11. Consider a single unit in the case envisaged in Example 2.2. More generally, assume that T has a hazard rate function

$$r(t|\sigma) \equiv \mathbf{1}_{\{t<\sigma\}} r_1(t) + \mathbf{1}_{\{t\geq\sigma\}} r_2(t)$$

and put $R_i(s) \equiv \int_0^s r_i(t)dt$. If σ is a non-observable random time with a density $q(\sigma)$, then the (predictive) survival function of T is

$$\overline{F}(s) = \int_0^s \exp\{-R_1(\sigma)\} \exp\{R_2(\sigma) - R_2(s)\} q(\sigma)d\sigma$$

$$+ \exp\{-R_1(s) \int_s^\infty q(\sigma)d\sigma\}. \tag{2.7}$$

One can show (see Exercise 2.65) that the predictive hazard failure rate is

$$r(t) = \alpha(t)r_1(t) + [1 - \alpha(t)]r_2(t)$$

where

$$\alpha(t) = \frac{\exp\{-R_1(t)\} \int_t^\infty q(\sigma)d\sigma}{\overline{F}(t)} = P\{\sigma > t | T > t\}.$$

Remark 2.12. As (2.2) and (2.3) show, a one-dimensional distribution for a lifetime T, admitting a density, can be equivalently characterized in terms of the distribution function, or the survival function or its hazard rate function.

In the next sections we shall consider multivariate analogues of arguments and formulae above.

2.2 Positive exchangeable random quantities

In this section we deal with some concepts related to joint distributions of several non-negative random quantities. Our treatment will in particular prepare the ground for the study of multivariate conditional hazard rate functions to be carried out in the next section.

Even though in the literature such concepts are introduced dealing with general cases, here we shall introduce them by directly thinking of exchangeable quantities $T_1, ..., T_n$, which represent the lifetimes of similar individuals $U_1, ..., U_n$.

If not otherwise stated, we deal with distributions on $\mathbb{R}_+^n$ admitting continuous joint density functions; a density function for n lifetimes will be denoted by the symbol $f^{(n)}(t_1, ..., t_n)$ or $f^{(n)}(\mathbf{t})$.

Illustrative examples will be presented at the end of the section.

A natural way to describe the joint distribution of $T_1, ..., T_n$ is by means of their joint (predictive) survival function:

$$\overline{F}^{(n)}(s_1, ..., s_n) \equiv P\{T_1 > s_1, ..., T_n > s_n\}$$

which often turns out to be more useful than the density function for the probabilistic modelling of the behavior of $U_1, ..., U_n$ and of our state of information about them.

Of course, $\overline{F}^{(n)}$ can be defined for any n-tuple of random quantities; in our case, $T_1, ..., T_n$ being exchangeable, we have that the function $\overline{F}^{(n)}$ is permutation-invariant.

It can be convenient to employ, in general, different notation for the arguments of $f^{(n)}$ and $\overline{F}^{(n)}$, for which we use the symbols $t_1, ..., t_n$ and $s_1, ..., s_n$, respectively: $t_1, ..., t_n$ are interpreted as *failure times*, while $s_1, ..., s_n$ are interpreted as *ages*.

Denoting by $F^{(n)}(s_1, ..., s_n)$ the joint distribution function, i.e. $F^{(n)}(s_1, ..., s_n) \equiv P\{T_1 \leq s_1, ..., T_n \leq s_n\}$, it is, trivially

$$\overline{F}^{(n)}(s_1, ..., s_n) + F^{(n)}(s_1, ..., s_n) \leq 1$$

(see Figure 3, for the case $n = 2$).

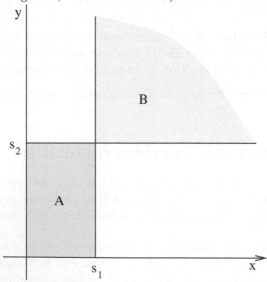

Figure 3. $F^{(2)}(s_1, s_2) = P\{\mathbf{T} \in A\}, \overline{F}^{(2)}(s_1, s_2) = P\{\mathbf{T} \in B\}$.

(Only for $n = 1$ it is $\overline{F}^{(1)}(s) + F^{(1)}(s) = 1, \forall s > 0$).
Moreover

$$f^{(n)}(t_1, ..., t_n) = (-1)^n \frac{\partial^n \overline{F}^{(n)}(s_1, ..., s_n)}{\partial s_1...\partial s_n}|_{s=t} \qquad (2.8)$$

which is in fact equivalent to

$$\overline{F}^{(n)}(s_1, ..., s_n) = \int_{s_1}^{\infty} ... \int_{s_n}^{\infty} f^{(n)}(t_1, ..., t_n)dt_1...dt_n \qquad (2.9)$$

For $1 \le k < n$, the marginal survival function of k quantities drawn from $T_1, ..., T_n$ is

$$\overline{F}^{(k)}(s_1, ..., s_k) \equiv P\{T_1 > s_1, ..., T_k > s_k\} = \overline{F}^{(n)}(s_1, ..., s_k, 0, ..., 0) \qquad (2.10)$$

Remark 2.13. In the case of i.i.d. random quantities with one-dimensional hazard rate function $r(t)$ and cumulative hazard function $R(t)$, the joint survival function can be written as

$$\overline{F}^{(n)}(s_1, ..., s_n) = \exp\{-\sum_{j=1}^{n} R(s_j)\} \qquad (2.11)$$

and their joint density function is

$$f^{(n)}(t_1, ..., t_n) = r(t_1)...r(t_n)\exp\{-\sum_{j=1}^{n} R(t_j) \qquad (2.12)$$

Following the discussion sketched in the previous chapter, it is natural in the predictive approach to consider the conditional distribution of unobserved variables, given the values taken by those which have been observed.

In the present setting we are often interested in computing *survival* probabilities for lifetimes, conditional on the values taken by already observed lifetimes. That is, we can be interested in conditional probabilities of the type

$$P\{T_{h+1} > s_{h+1}, ..., T_n > s_n | T_1 = t_1, ..., T_h = t_h\}. \qquad (2.13)$$

Taking into account the equation (2.9) and adapting (1.37) of Chapter 1, we obtain

$$P\{T_{h+1} > s_{h+1}, ..., T_n > s_n | T_1 = t_1, ..., T_h = t_h\} =$$

$$\int_{s_{h+1}}^{\infty} ... \int_{s_n}^{\infty} f^{(n-h)}(t_{h+1}, ..., t_n | T_1 = t_1, ..., T_h = t_h)dt_{h+1}...dt_n =$$

$$\frac{\int_{s_1}^{\infty} ... \int_{s_n}^{\infty} f^{(n)}(t_1, ..., t_h, t_{h+1}, ..., t_n)dt_{h+1}...dt_n}{f^{(h)}(t_1, ..., t_h)} \qquad (2.14)$$

Remark 2.14. The existence of a joint density $f^{(n)}$ allows us to give a precise meaning to the formal expression

$$P\{T_{h+1} > s_{h+1}, ..., T_n > s_n | T_1 = t_1, ..., T_h = t_h\}$$

even if $\{T_1 = t_1, ..., T_h = t_h\}$ is an event of null probability. Indeed we implicitly define

$$P\{T_{h+1} > s_{h+1}, ..., T_n > s_n | T_1 = t_1, ..., T_h = t_h\} \equiv$$

$$\lim_{\delta \to 0} P\{T_{h+1} > s_{h+1}, ..., T_n > s_n | t_1 \leq T_1 \leq t_1 + \delta, ..., t_h \leq T_h \leq t_h + \delta\} \quad (2.15)$$

Under absolute continuity the above limit exists and Equation (2.14) provides an expression for it.

Note that, in the statistical analysis of lifetime data, an event of the type $\{T_1 = t_1, ..., T_h = t_h\}$ corresponds to the observation of h failures and $t_1, ..., t_h$ denote *failure times*.

Commonly, one also observes *survival data*. That is, one fairly often must also take into account previous pieces of information about survival of some of the individuals $U_1, ..., U_n$ and, rather than in conditional survival probability of the form above, we are interested, for $s_1 \geq r_1, ..., s_n \geq r_n \geq 0$, in

$$P\{T_1 > s_1, ..., T_n > s_n | T_1 > r_1, ..., T_n > r_n\} \quad (2.16)$$

or

$$P\{T_{h+1} > s_{h+1}, ..., T_n > s_n | D\} \quad (2.17)$$

where

$$D \equiv \{T_1 = t_1, ..., T_h = t_h, T_{h+1} > r_{h+1}, ..., T_n > r_n\}. \quad (2.18)$$

Of course some of the r_i $(1 \leq i \leq n)$ can be equal to zero, and, in particular (2.17) reduces to (2.13) when $r_i = 0$ $(h + 1 \leq i \leq n)$.

As a function of the variables $\tau_1 = s_1 - r_1, ..., \tau_n = s_n - r_n$, the conditional probability in (2.16) can then be seen to be the joint survival functions of the residual lifetimes $T_j - r_j$ $(r = h + 1, ..., n)$ conditional on the observation of survival data. Trivially, we can write

$$P\{T_1 > s_1, ..., T_n > s_n | T_1 > r_1, ..., T_n > r_n\} = \frac{\overline{F}^{(n)}(s_1, ..., s_n)}{\overline{F}^{(n)}(r_1, ..., r_n)} \quad (2.19)$$

The conditional probability in (2.17) can be defined as

$$\lim_{\delta \to 0} P\{T_{h+1} > r_{h+1} + \tau_{h+1}, ..., T_n > r_n + \tau_n | D_\delta\}$$

where

$$D_\delta \equiv \{t_1 \leq T_1 \leq t_1 + \delta, ..., t_h \leq T_h \leq t_h + \delta, T_{h+1} > r_{h+1}, ..., T_n > r_n\}$$

Again it can be seen as the joint survival function of residual lifetimes of surviving individuals, conditional on the observation of survival data and/or failure data. An explicit expression for it, in terms of $f^{(n)}$, is provided by the following

Proposition 2.15. *For* $h = 1, ..., n - 1$, $t_j \geq 0$ $(0 \leq j \leq h)$, $0 \leq r_j \leq s_j$ $(h + 1 \leq j \leq n)$, *one has*

$$P\{T_{h+1} > s_{h+1}, ..., T_n > s_n | D\} =$$

$$\frac{\int_{s_{h+1}}^{\infty} \cdots \int_{s_n}^{\infty} f^{(n)}(t_1, ..., t_h, t_{h+1}, ..., t_n) dt_{h+1} ... dt_n}{\int_{r_{h+1}}^{\infty} \cdots \int_{r_n}^{\infty} f^{(n)}(t_1, ..., t_h, t_{h+1}, ..., t_n) dt_{h+1} ... dt_n} \tag{2.20}$$

Proof. Since $r_j \leq s_j$ $(h + 1 \leq j \leq n)$, the event $\{T_{h+1} > s_{h+1}, ..., T_n > s_n\}$ implies $\{T_{h+1} > r_{h+1}, ..., T_n > r_n\}$.
Thus

$$\lim_{\delta \to 0} P\{T_{h+1} > s_{h+1}, ..., T_n > s_n | D_\delta\} =$$

$$\lim_{\delta \to 0} \frac{P\{T_{h+1} > s_{h+1}, ..., T_n > s_n | t_1 \leq T_1 \leq t_1 + \delta, ..., t_h \leq T_h \leq t_h + \delta\}}{P\{T_{h+1} > r_{h+1}, ..., T_n > r_n | t_1 \leq T_1 \leq t_1 + \delta, ..., t_h \leq T_h \leq t_h + \delta\}}$$

$$= \frac{P\{T_{h+1} > s_{h+1}, ..., T_n > s_n | T_1 = t_1, ..., T_h = t_h\}}{P\{T_{h+1} > r_{h+1}, ..., T_n > r_n | T_1 = t_1, ..., T_h = t_h\}}$$

whence (2.20) is obtained by applying (2.14).   □

Now we want to specialize the above arguments to common special cases of infinite extendibility. More precisely we shall now consider parametric models, as defined by the condition (1.38) in Section 1.3.
Let us then assume that the lifetimes $T_1, ..., T_n$ are conditionally i.i.d., given a parameter Θ taking values in $L \subset \mathbb{R}^d$ for some $d \geq 1$, with an initial density π_0, and that the one-dimensional conditional survival function $\overline{G}(t|\theta)$ admits a density $g(t|\theta)$ $(\forall \theta \in L)$. We denote by $r(t|\theta)$ and $R(s|\theta)$ the hazard and cumulative hazard functions, respectively, corresponding to $\overline{G}(t|\theta)$.

By recalling the equations (2.11) and (2.12), we have

$$\overline{F}^{(n)}(s_1, ..., s_n) = \int_L \overline{G}(s_1|\theta)...\overline{G}(s_n|\theta)\pi_0(\theta)d\theta =$$

$$\int_L \exp\{-\sum_{j=1}^n R(s_j|\theta)\}\pi_0(\theta)d\theta, \tag{2.21}$$

$$f^{(n)}(t_1, ..., t_n) = \int_L g(t_1|\theta)...g(t_n|\theta)\pi_0(\theta)d\theta = \tag{2.22}$$

$$\int_L r(t_1|\theta)...r(t_n|\theta) \cdot \exp\left\{-\sum_{j=1}^n R(t_j|\theta)\right\}\pi_0(\theta)d\theta.$$

Under the assumptions on $g(\cdot|\theta)$ which allow interchanging the integration order, by Proposition 2.15, we can then write, for D as in (2.18),

$$P\{T_{h+1} > s_{h+1}, ..., T_n > s_n|D\} =$$

$$\frac{\int_L \int_{s_{h+1}}^\infty \cdots \int_{s_n}^\infty g(t_1|\theta)...g(t_h|\theta)g(t_{h+1}|\theta)...g(t_n|\theta)\pi_0(\theta)dt_{h+1}...dt_n d\theta}{\int_L \int_{r_{h+1}}^\infty \cdots \int_{r_n}^\infty g(t_1|\theta)...g(t_h|\theta)g(t_{h+1}|\theta)...g(t_n|\theta)\pi_0(\theta)dt_{h+1}...dt_n d\theta}. \tag{2.23}$$

Consider now the *posterior* (i.e. conditional) density of Θ given the observation D. In order to derive it, we can use Bayes' formula two times subsequently: the first time to account for failure data $\{T_1 = t_1, ..., T_h = t_h\}$ and the second time to account for survival data $\{T_{h+1} > r_{h+1}, ..., T_n > r_n\}$. This gives

$$\pi(\theta|D) = \frac{g(t_1|\theta)...g(t_h|\theta)\overline{G}(r_{h+1}|\theta)...\overline{G}(r_n|\theta)\pi_0(\theta)}{\int_L g(t_1|\theta)...g(t_h|\theta)\overline{G}(r_{h+1}|\theta)...\overline{G}(r_n|\theta)\pi_0(\theta)d\theta} \tag{2.24}$$

Equation (2.24) shows that the ratio in the r.h.s. of (2.23) can be rewritten as

$$\int_L \prod_{j=h+1}^n \frac{\overline{G}(s_j|\theta)}{\overline{G}(r_j|\theta)} \times \pi(\theta|D)\, d\theta \tag{2.25}$$

and we can conclude with the following

Proposition 2.16. *Assume $\overline{F}^{(n)}(s_1, ..., s_n)$ to be of the form (2.21). Then it is*

$$P\{T_{h+1} > s_{h+1}, ..., T_n > s_n | D\} =$$

$$\int_L \frac{P\{T_{h+1} > s_{h+1}, ..., T_n > s_n | \theta\}}{P\{T_{h+1} > r_{h+1}, ..., T_n > r_n | \theta\}} \times \pi(\theta | D) \, d\theta \qquad (2.26)$$

Notice that (2.26) can be seen as a natural multivariate extension of (2.4).

In order to illustrate the arguments introduced so far, let us now turn to discussing some of the special cases introduced in the last section.

Example 2.17. (Finite forms of time transformed exponential models). For $s \geq 0$, denote by $\Sigma_s^{(n)}$ the simplex defined by

$$\Sigma_s^{(n)} \equiv \{ \mathbf{t} \in \mathbb{R}_+^n \mid \sum_{i=1}^n t_i = s \}$$

and, for n lifetimes $T_1, ..., T_n$, we consider the (singular) probability distribution which is uniform over $\Sigma_s^{(n)}$; the joint survival function of $(T_1, ..., T_n)$ is in this case given by

$$\overline{F}^{(n)}(s_1, ..., s_n) = \begin{cases} \left(1 - \frac{\sum_{i=1}^n s_i}{s}\right)^{n-1} & \text{for } 0 \leq \sum_{i=1}^n s_i \leq s \\ 0 & \text{otherwise} \end{cases} ;$$

we shall also write this as

$$\overline{F}^{(n)}(s_1, ..., s_n) = \left[1 - \frac{\sum_{i=1}^n s_i}{s}\right]_+^{n-1} .$$

See Figure 4 for the case $n = 2$.

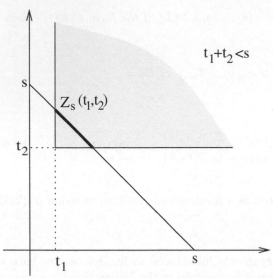

Figure 4 The bold segment $Z_s(t_1, t_2)$ has length $(1 - t_1 - t_2)\sqrt{2}$ and Σ_s has length $s\sqrt{2}$. $P\{T_1 > t_1, T_2 > t_2\} = P\{(T_1, T_2) \in Z_s(t_1, t_2)\} = \frac{(1-t_1-t_2)\sqrt{2}}{s\sqrt{2}} = \left[1 - \frac{t_1+t_2}{s}\right]_+$.

Let $Z_1, ..., Z_n$ be i.i.d. exponential lifetimes and set $S_n = \sum_{i=1}^n Z_i$; for $s > 0$, the conditional distribution of $Z_1, ..., Z_n$ given $\{S_n = s\}$ is the uniform distribution over $\Sigma_s^{(n)}$. This same fact remains true for whatever distribution of $Z_1, ..., Z_n$ satisfying the condition

$$P\{Z_1 > s_1, ..., Z_n > s_n\} = \overline{\Phi}\left(\sum_{i=1}^n s_i\right)$$

i.e. such that the joint survival function is Schur-constant.

This argument shows that, conditioning with respect to S_n and denoting by $P_{\overline{\Phi}}^{(n)}$ the marginal distribution of S_n, we can write, for an arbitrary Schur-constant survival function $\overline{F}^{(n)}$,

$$\overline{F}^{(n)}(\mathbf{s}) = \overline{\Phi}\left(\sum_{i=1}^n s_i\right) = \int_0^\infty \left[1 - \frac{\sum_{i=1}^n s_i}{s}\right]_+^{n-1} dP_{\overline{\Phi}}^{(n)}(s).$$

$Z_1, ..., Z_n$ are i.i.d. exponential (i.e. $\overline{\Phi}(t) = \exp\{-\theta t\}$, for some $\theta > 0$) if and only if $P_{\overline{\Phi}}^{(n)}$ is $G(n, \theta)$, a gamma distribution with parameters n and θ.

$Z_1, ..., Z_n$ are conditionally i.i.d. exponential given a parameter Θ, if and only if $P_{\overline{\Phi}}^{(n)}$ is a mixture of gamma distributions with parameters n and θ,

$\{G(n, \theta)\}_{\theta \geq 0}$. This happens if and only if the function $\overline{\Phi}$ is such that the identity $\overline{F}^{(n)}(\mathbf{s}) = \overline{\Phi}\left(\sum_{i=1}^{n} s_i\right)$ gives rise to a well-defined n-dimensional survival function, for any $n = 1, 2, \ldots$.

We can also state the latter arguments in the following form:

for $n \to \infty$, a uniform distribution over the simplex $\Sigma_s^{(n)}$ tends to coincide with the distribution of n i.i.d lifetimes.

This is the more general formulation of the de Finetti-type Theorem 1.44 proven by Diaconis and Freedman (1987).

For $Z_1, \ldots, Z_n$ with a Schur-constant survival function, consider now lifetimes $T_1, \ldots, T_n$ of the form

$$T_1 = R^{-1}(Z_1), \ldots, T_n = R^{-1}(Z_n)$$

where R is a strictly increasing non-negative function. We have

$$\overline{F}_{\mathbf{T}}^{(n)}(\mathbf{s}) = P\{Z_1 > R(s_1), \ldots, Z_n > R(s_n)\} = \overline{\Phi}\left(\sum_{i=1}^{n} R(s_i)\right).$$

We call this a finite form of time transformed exponential model (see also Barlow and Mendel, 1992).

Proportional hazard models which will be discussed in the next example can be obtained from these models in terms of the de Finetti-type result mentioned above.

Example 2.18. (Proportional hazards, conditionally i.i.d. lifetimes with a time transformed exponential model). Let Θ be a non-negative parameter with a prior density $\pi_0(\theta)$ and let $T_1, \ldots, T_n$ be conditionally i.i.d given Θ, with a conditional distribution of the form

$$\overline{G}(s|\theta) = \exp\{-\theta R(s)\} \tag{2.27}$$

where $R : [0, +\infty) \to [0, +\infty)$ is a non-decreasing differentiable function such that $R(0) = 0$ and $\lim_{s \to \infty} R(s) = +\infty$.

By denoting $\rho(t) \equiv \frac{d}{ds}R(s)|_{s=t}$, Equations (2.21) and (2.22), respectively, become

$$\overline{F}^{(n)}(s_1, \ldots, s_n) = \int_0^{+\infty} \exp\{-\theta \sum_{j=1}^{n} R(s_j)\}\pi_0(\theta)d\theta \tag{2.28}$$

$$f^{(n)}(t_1, \ldots, t_n) = \int_0^{+\infty} \theta^n \prod_{j=1}^{n} \rho(t_j) \times \exp\{-\theta \sum_{j=1}^{n} R(t_j)\}\pi_0(\theta)d\theta. \tag{2.29}$$

It follows, for D as in (2.18), that

$$\pi(\theta|D) = k\,(h; \mathbf{t}; \mathbf{r}) \times \theta^h \exp\{-\theta[\sum_{j=1}^{h} R(t_j) + \sum_{j=h+1}^{n} R(r_j)]\}\pi_0(\theta) \qquad (2.30)$$

where $k\,(h; \mathbf{t}; \mathbf{r})$ is the normalization constant. Whence, using (2.26), we obtain

$$P\{T_{h+1} > s_{h+1}, ..., T_n > s_n|D\} =$$

$$\frac{\int_0^\infty \theta^h \exp\{-\theta[\sum_{j=1}^{h} R(t_j) + \sum_{j=h+1}^{n} R(s_j)]\}\pi_0(\theta)d\theta}{\int_0^\infty \theta^h \exp\{-\theta[\sum_{j=1}^{h} R(t_j) + \sum_{j=h+1}^{n} R(r_j)]\}\pi_0(\theta)d\theta} \qquad (2.31)$$

It is worth remarking that $\overline{F}^{(n)}(s_1, ..., s_n)$ actually is a function of $\sum_{j=1}^{n} R(s_j)$; furthermore the above conditional probability depends upon the quantities

$$(t_1, ..., t_h, r_{h+1}, ..., r_n)$$

only through the triple

$$\left(h, \sum_{j=1}^{h} R(t_j), \sum_{j=h+1}^{n} R(r_j)\right)$$

and it depends upon $(s_{h+1}, ..., s_n)$ only through the quantity $\sum_{j=h+1}^{n} R(s_j)$.

In particular, in the conditionally exponential case ($R(s) \equiv s$), it depends upon $(t_1, ..., t_n, r_{h+1}, ..., r_n)$ only through the pair $(h, \sum_{j=1}^{h} t_j + \sum_{j=h+1}^{n} r_j)$.

Example 2.19. Continuing Example 2.7, we now consider two lifetimes T_1, T_2, conditionally independent given the random time σ, with hazard rate function

$$r(t|\sigma) \equiv \mathbf{1}_{\{t<\sigma\}}r_1\,(t) + \mathbf{1}_{\{t\geq\sigma\}}r_2\,(t)$$

$q(\cdot)$ denotes the density of σ and $R_i(t) = \int_0^t r_i(u)du$. In order to give $\overline{F}^{(2)}(s_1, s_2)$ an explicit form, it is convenient to write

$$\overline{F}^{(2)}(s_1, s_2) = \int_0^{s_1} P\{T_1 > s_1, T_2 > s_2|\sigma\}q(\sigma)d\sigma +$$

$$\int_{s_1}^{s_2} P\{T_1 > s_1, T_2 > s_2|\sigma\}q(\sigma)d\sigma + \int_{s_2}^{\infty} P\{T_1 > s_1, T_2 > s_2|\sigma\}q(\sigma)d\sigma.$$

Whence, it can be checked (Exercise 2.67) that

$$P\{T_1 > s_1, T_2 > s_2|T_1 > r, T_2 > r\} = \frac{\overline{F}^{(2)}(s_1, s_2)}{\overline{F}^{(2)}(r, r)}$$

$$= \int_0^\infty P\{T_1 > s_1, T_2 > s_2 | \sigma\} q(\sigma | T_1 > r, T_2 > r) d\sigma. \qquad (2.32)$$

Example 2.20. (Heterogeneous populations). Consider the case of Example 2.3 and let the distribution G_i be characterized by a density function $g_i(t)$ and by a one-dimensional hazard rate

$$r_i(t) = g_i(t) / \overline{G}_i(t), (i = 0, 1).$$

Furthermore let $p_0(m) = P\{M = m\}$ $(m = 0, 1, ..., n)$ denote the prior distribution of the total number M of weak individuals.

The joint survival function and the joint density function of $T_1, ..., T_n$ are easily seen to have, respectively, the form

$$\overline{F}^{(n)}(s_1, ..., s_n) = \sum_{m=0}^n p_0(m) P\{T_1 > s_1, ..., T_n > s_n | M = m\} =$$

$$\sum_{m=0}^n p_0(m) \frac{1}{n!} \sum_\pi \prod_{i=1}^m \overline{G}_1(s_{\pi_i}) \prod_{i=m+1}^n \overline{G}_0(s_{\pi_i}), \qquad (2.33)$$

$$f^{(n)}(t_1, ..., t_n) =$$

$$\sum_{m=0}^n p_0(m) \frac{1}{n!} \sum_\pi \prod_{i=1}^m r_1(s_{\pi_i}) \overline{G}_1(s_{\pi_i}) \prod_{i=m+1}^n r_0(s_{\pi_i}) \overline{G}_0(s_{\pi_i}),$$

where the sum is taken over all the permutations $\pi = (\pi_1, ..., \pi_n)$ of $(1, 2, ..., n)$.
The marginal one-dimensional survival function of a lifetime T_i is

$$\overline{F}^{(1)}(s) = \overline{F}^{(n)}(s, 0, 0, ..., 0) = \overline{G}_1(s) \frac{E(M)}{n} + \overline{G}_0(t) \frac{n - E(M)}{n}$$

and the marginal one-dimensional density function is

$$f^{(1)}(t) = g_1(t) \frac{E(M)}{n} + g_0(t) \frac{n - E(M)}{n}$$

As far as the marginal hazard rate function is concerned, we have

$$\lambda(t) = \frac{f^{(1)}(t)}{\overline{F}^{(1)}(t)} = \alpha(t) r_1(t) + [1 - \alpha(t)] r_0(t)$$

where

$$\alpha(t) = \frac{\overline{G}_1(t) E[M]}{\overline{G}_1(t) E[M] + \overline{G}_0(t)(n - E[M])}$$

It can be easily shown that $\alpha(t)$ has the meaning

$$\alpha(t) = P(C_j = 1|T_j > t).$$

As (2.33) shows, the prior distribution of M of course influences the joint distribution of $T_1, ..., T_n$.

We consider now the conditional survival probabilities, given failure and survival data; for the sake of notation simplicity, we limit ourselves to the case $n = 2$.

The conditional survival probability $P\{T_2 > s_2|T_1 = t_1, T_2 > r_2\}$ can be obtained by specializing the equation (2.20). It can be also obtained by noticing that, for $D \equiv \{T_1 = t_1, T_2 > r_2\}$, it must be

$$P\{T_2 > s_2|D\} =$$

$$P\{T_2 > s_2|T_2 > r_2, K_1 = 1, K_2 = 1\} \times P\{K_1 = 1, K_2 = 1|D\}+$$

$$P\{T_2 > s_2|T_2 > r_2, K_1 = 1, K_2 = 0\} \times P\{K_1 = 1, K_2 = 0|D\} +$$

$$P\{T_2 > s_2|T_2 > r_2, K_1 = 0, K_2 = 1\} \times P\{K_1 = 0, K_2 = 1|D\}+$$

$$P\{T_2 > s_2|T_2 > r_2, K_1 = 0, C_2 = 0\} \times P\{K_1 = 0, K_2 = 0|D\}.$$

In the special case $M_0 \sim b(2, p)$, where as already remarked T_1, T_2 are stochastically independent, the above reduces to

$$P\{T_2 > s_2|D\} = P\{T_2 > s_2|T_2 > r_2, K_2 = 1\}P\{K_2 = 1|T_2 > r_2\}+$$

$$P\{T_2 > s_2|T_2 > r_2, K_2 = 0\}P\{K_2 = 0|T_2 > r_2\}.$$

Remark 2.21. ("Conditional formula of total probabilities"). A fundamental property of conditional probability specializes in the following formula. Let E and D be two events with positive probabilities and let Θ be a random quantity taking values in a (one-dimensional or multidimensional) space L. Then one has

$$P(E|D) = \int_L P(E|D; \Theta = \theta)d\Pi(\theta|D) \qquad (2.34)$$

where $\Pi(\theta|D)$ denotes the conditional distribution of Θ, given D. This formula has a great importance in our setting and the examples above show the following: computations of conditional survival probabilities given D, with D being a survival data of the type $\{T_1 > r_1, ..., T_n > r_n\}$ or, more in general, of the form (2.18), can just be recognized as different applications of this formula, taking as Θ a suitable "latent" variable.

Recall that, when we write $P(E|C)$, C being an event of null probability we tacitly assume that $P(E|C)$ can be properly defined as a limit of "well-defined" conditional probabilities similarly to what we did in Remark 2.14.

Example 2.22. (Common mode failures). Consider two units which are similar and work independently one of the other; they are, however, sensible to the same destructive shock. We assume that, in the absence of a shock, the two lifetimes would be independent with the same survival function $\overline{G}$ and that the waiting time until the shock is W, where the survival function of W is $\overline{H}$. We can then write

$$T_1 = \min(V_1, W), T_2 = \min(V_2, W)$$

where V_1, V_2 are independent lifetimes with survival function $\overline{G}$. For the joint survival function we have then

$$\overline{F}^{(2)}(s_1, s_2) = P\{V_1 > s_1, V_2 > s_2, W > s_1, W > s_2\} =$$

$$\overline{G}(s_1)\,\overline{G}(s_2)\,\overline{H}(s_1 \vee s_2)$$

where $s_1 \vee s_2$ is the short-hand notation for $\max(s_1, s_2)$.

Note that, since $P\{T_1 = T_2\} > 0$, $\overline{F}^{(2)}(s_1, s_2)$ does not admit a joint density; then we cannot use the formula (2.20) to compute a conditional survival probability of the type $P\{T_2 > s_2 | T_1 = t_1, T_2 > r_2\}$.

However a simple reasoning yields, for $s_2 > r_2 > t_1$,

$$P\{T_2 > s_2 | T_1 = t_1, T_2 > r_2\} = \frac{\overline{G}(s_2)\,\overline{H}(s_2)}{\overline{G}(r_2)\,\overline{H}(r_2)}, \tag{2.35}$$

and, for $r_2 < t_1$,

$$P\{T_2 > s_2 | T_1 = t_1, T_2 > r_2\} = \frac{\overline{G}(s_2)}{\overline{G}(r_2)}\overline{H}(s_2)\,\frac{g(t_1)}{\overline{H}(t_1)\,g(t_1) + \overline{G}(t_1)h(t_1)}$$

Example 2.23. (Bivariate model of Marshall-Olkin). Consider again the model in the previous example, assuming in particular that the survival functions $\overline{G}$ and $\overline{H}$ are exponential, $\overline{G}(s) = \exp\{-\lambda s\}$, $\overline{H}(s) = \exp\{-\lambda' s\}$ say. In this special case we obtain

$$\overline{F}^{(2)}(s_1, s_2) = \exp\{-\lambda(s_1 + s_2) - \lambda'(s_1 \vee s_2)\} \tag{2.36}$$

This is known as the bivariate exponential model of Marshall-Olkin (Marshall and Olkin, 1967). Notice that the one-dimensional marginal distribution is exponential.

Remark 2.24. In the reliability practice, data are often generated according to a scheme as follows. Think for instance of the case of n similar devices which are available to be installed into m different positions ($m < n$). At time 0, m new devices are installed; at the instant of a failure, the failed device is replaced by a new one in the same position. At a generic instant $t > 0$, the observed data can be represented in the form

$$D \equiv \{T_1 = t_1, ... T_h = t_h, T_{h+1} > r_1, ..., T_{h+m} > r_m, T_{h+m+1} > 0, ..., T_n > 0\}$$

where h is the number of replacements that have occurred up to t, $t_1, ..., t_h$ are the observed lifetimes for the corresponding devices and $0 < r_i \le t$ ($i = 1, ..., m$). r_i is the age of the device working in the position i at time t, and $r_i = t$ means that no replacement previously occurred in that position. In this scheme it is, at any instant t,

$$\sum_{j=1}^{h} t_j + \sum_{j=1}^{m} r_j = mt.$$

As special cases of the above scheme, we encounter two special patterns of observed data.

(A) One special case is when $m = 1$; in such a case then, obviously, only one unit is working at a time and then there is at most only one survival data, at any instant t.

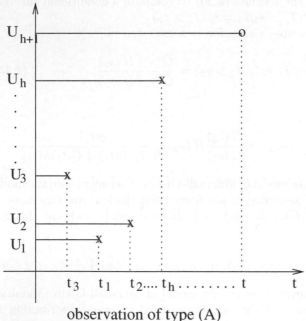

observation of type (A)

Figure 5.1.

(B) A second pattern of failure and survival data is the one that we obtain from the above scheme when no replacement is scheduled for failed devices. This means that all the individuals start living, or operating, at the same time and that all of those surviving at a time $t > 0$ share the same age t $(m = n)$. We then consider a duration experiment where the units $U_1, ..., U_n$ are new at time 0 and are simultaneously put under test, progressively recording all the subsequent failure times. At any time $t > 0$, the available statistical observation then is a history of the form

$$D' \equiv \{T_{i_1} = t_1, ..., T_{i_h} = t_h, T_{j_1} > t, ..., T_{j_{n-h}} > t\} \qquad (2.37)$$

where $0 \leq h \leq n$, $0 < t_1 \leq ... \leq t_h \leq t$, $i_1, ..., i_h \subset \{1, 2, ..., n\}$ is the set of the indexes denoting the units which failed up to t and

$$\{j_1, ..., j_{n-h}\} \equiv \{1, 2, ..., n\} \setminus \{i_1, ..., i_h\}$$

is the set of the indexes denoting the units which are still surviving at time t.

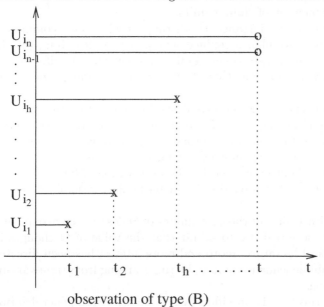

observation of type (B)

Figure 5.2.

This is often described in the literature as the case of *longitudinal observations*. A *dynamic approach* is appropriate to deal with such situations. This means in particular that the form of interdependence existing among lifetimes is more conveniently described in terms of the notion of multivariate conditional hazard rates, which will be introduced in the next section.

It is to be noticed, on the other hand, that different kinds of patterns for failure and survival data observations are often encountered in survival analysis, where data from the biomedical field are to be considered.

Remark 2.25. Survival data typically come along with the presence of some rule for stopping the life-data observation. In some cases the stopping rule can itself be *informative,* i.e. it can contain some information relatively to the lifetimes we want to predict (see e.g. Barlow and Proschan, 1988). In the derivation of Proposition 2.15, which is the basis for the subsequent developments, we tacitly assumed that the stopping rule is not informative.

Parametric models described by (2.21) are often encountered in applications. As mentioned in the previous section, the (finite-dimensional) parameter Θ can be interpreted as a vector specifying factors (such as structural properties, environmental conditions, or resistance to stress situations) which simultaneously affect the behavior of the individuals $U_1, ..., U_n$.

This is a special case of "infinite extendibility." Note that this latter condition is a natural assumption when we think of units which come from a conceivably infinite population of similar units.

In the reliability practice, for instance, this is the case when, for a component in a position of a system, we plan a replacement policy by installing (at the instant of failure or at a fixed age) other similar components: the lifetimes of new components, which are subsequently put into operation, form a conceivably infinite population.

Conditional independence in (2.21) formalizes the assumption that there is no "physical interaction" among $U_1, ..., U_n$ and, at the same time, that their behavior is influenced by the common factors specified by Θ. This assumption entails that the stochastic dependence among $T_1, ..., T_n$, is only due to the increase of information, about Θ, carried by the observation of failures or survivals of some of the U's. Actually, this is, in particular, formalized by the formula (2.25) above.

However, when at least some of the coordinates of Θ describe environmental conditions, it may be more realistic to admit that the value of Θ changes in time. In recent statistical literature, more and more interest has been directed toward the study of interdependence among lifetimes arising from time-varying environmental conditions.

A simple model of this type is provided by a straightforward generalization of the *proportional hazard model* of Example 2.18: the parameter Θ is replaced by a stochastic process η_t with state space contained in $\mathbb{R}_+$ and (conditional on the knowledge of η_t) the hazard rate of each lifetime, at time $t > 0$, is modeled to be $\eta_t \times r(t)$, where $r(t)$ is a given hazard rate function.

This model can be interpreted as follows: under ideal, laboratory conditions, each lifetime has a baseline hazard rate $r(t)$ whereas actual operating environment causes $r(t)$ to be modulated by the *environmental factor function* η_t (see the review paper by Singpurwalla, 1995).

In a Bayesian viewpoint, the assessment of initial conditions and of the law for the stochastic process η_t here replaces the prior distribution assessed for a time-constant parameter Θ.

Of course different kinds of multivariate survival functions (and then of interdependence among $T_1, ..., T_n$) are obtained under different models for the evolution of η_t. In some particular cases, the laws of the process η_t have special structures (see e.g. the models in the examples at the end of this section) which allow one to derive the joint (predictive) survival function.

In general it is not at all feasible obtaining the joint survival function in a closed form and it is only possible to write down "dynamic" equations which define $\overline{F}^{(n)}$ in an implicit way.

Furthermore an even more complicated situation is in general to be taken into account: the conditional hazard rate of any individual at time t may depend not only on the environmental conditions at the same time t but also on their "past history" and/or on the history of failures and survivals

$$D \equiv \{T_1 = t_1, ..., T_h = t_h, T_{h+1} > r_{h+1}, ..., T_n > r_n\}$$

observed up to t.

Apart from specific probabilistic models, the mathematical tools to deal with general problems involving time-varying environmental conditions is to be found in the framework of the theory of point processes and of stochastic filtering (Bremaud, 1981; Arjas, 1981).

Remark 2.26. Assuming a common dynamic environment presupposes the consideration of units which can work simultaneously, as for instance happens in the situations described in the previous Remark 2.24. However this is in general a rather complicated situation to deal with and one usually restricts attention to the case where dynamic histories are observed of the type in (2.37); i.e., in the case of possibly dynamic environment, one usually considers the case (B), where individuals start living, or operating, at the same time, so that all of those surviving at a time $t > 0$ share the same age t.

This is the case which will be considered in detail in the next section when introducing the concept of *multivariate conditional hazard rate* functions. Such a concept is in particular helpful to describe cases when, even conditionally on the knowledge of the environment process, the lifetimes are not necessarily independent.

The following two examples show special models obtained by specifying arguments contained in Singpurwalla and Yougren (1993) to the case of exchangeable lifetimes.

Example 2.27. Let $A(t) \equiv \int_0^t a(\xi)d\xi(t \geq 0)$ be a given non-decreasing left-continuous real valued function and let $b \in (0, +\infty)$. A *gamma process* with parameters $A(t)$ and $1/b$ is a stochastic process X_t characterized by the following conditions
  (i) $P\{X_0 = 0\} = 1$
  (ii) X_t has independent increments

(iii) $X_t - X_s$ has a gamma distribution with parameters

$$\alpha = A(t) - A(s), \beta = 1/b(s < t)$$

Consider now the case of two (conditionally) independent lifetimes T_1, T_2 with a continuous baseline hazard $r(t)$ and environmental factor function η_t, η_t being a gamma process with parameters $A(t)$ and $1/b$. Then the joint survival function of T_1, T_2 is given, for $s_1 \leq s_2$, by

$$\overline{F}^{(2)}(s_1, s_2) = \exp\left\{ -\int_0^{s_1} \log[1 + \frac{1}{b} \int_u^{s_1} 2r(\xi)d\xi]a(u)du \right\} \qquad (2.38)$$

$$\times \exp\left\{ -\int_{s_1}^{s_2} \log[1 + \frac{1}{b} \int_u^{s_2} r(\xi)d\xi]a(u)du \right\}$$

Example 2.28. Here we consider conditionally independent lifetimes with a constant baseline hazard r and an environmental factor function η_t of the form

$$\eta_t = \sum_{k=0}^{\infty} X_k \times h(t - \mathsf{T}_{(k)})$$

where $X_1, X_2, \ldots$ are non-negative i.i.d. random variables with a common distribution G, $\mathsf{T}_{(1)}, \mathsf{T}_{(2)}, \ldots$ are the arrival times of a Poisson process with a known rate $m(t)$ and h is a non-negative function such that $h(u) = 0$ for $u \leq 0$ (see e.g. Cox and Isham, 1980, p. 135). Furthermore assume that $X_1, X_2, \ldots$ are independent of $\mathsf{T}_{(1)}, \mathsf{T}_{(2)}, \ldots$. Let G^* be the Laplace transform of G, it can be proved (Singpurwalla and Yougren, 1993) that, for $s_1 \leq s_2$, the two-dimensional survival function is

$$\overline{F}^{(2)}(s_1, s_2) = G^*(r \times [H(s_1) + H(s_2)]) \times$$

$$\exp\left\{ \int_0^{s_1} G^*(r \times [H(s_1 - u_1) + H(s_2 - u_1)]) \times m(u_1)du_1 \right\} \times$$

$$\exp\left\{ \int_{s_1}^{s_2} G^*(r \times H(s_2 - u_2)) \times m(u_2)du_2 \right\} \qquad (2.39)$$

In the special case when $H(u) \equiv 1$, $m(u) \equiv m$, G is a gamma distribution with parameters α and $\beta = \frac{r}{m}$, then (2.39) becomes

$$\overline{F}^{(2)}(s_1, s_2) = \left(\frac{1 + m - (s_2 - s_1)}{1 + m(s_1 + s_2)} \right)^{\frac{1}{2}} \exp\{-m \times s_2\}.$$

2.3   Multivariate conditional hazard rates

In this section we assume the existence of a continuous joint density $f^{(n)}$ for lifetimes $T_1, ..., T_n$ and specifically consider the "longitudinal" observation of the behavior of the individuals $U_1, ..., U_n$; this means that the failure and survival data, observed in the interval $[0, t]$ are of the special form (2.37).

This situation leads, in a natural way, to the definition of the concept of multivariate conditional hazard rate (m.c.h.r.).

Such a concept can be seen as a direct extension of the one-dimensional hazard rate; it provides a tool for characterizing (absolutely continuous) distributions on $\mathbb{R}^n_+$, and is an alternative to traditional tools such as density functions or survival functions. In various cases it is a particularly handy method to describe a probabilistic model, since it conveys the "dynamic" character of the statistical observations in the applications we are dealing with.

Also in formulating the definition of m.c.h.r. we shall refer to the special case of exchangeability. More generally, this concept was considered e.g. by Gaver (1963), Lawless (1982), and extensively used later on by Shaked and Shanthikumar in the study of dynamic-type dependence and aging properties of vectors of lifetimes. We shall briefly report the general definition in the next Chapter. This is related to the more general concept of *stochastic intensity*, introduced in the framework of the theory of point processes (see Arjas, 1981; Bremaud, 1981).

If not otherwise specified, the individuals $U_1, ..., U_n$ will be thought of as n pieces of an industrial equipment, since such interpretation actually provides a more flexible basis for the language that we shall use.

Let $\overline{F}^{(n)}(s_1, ..., s_n)$ be the survival function of $T_1, ..., T_n$ and let $T_{(1)} \leq ... \leq T_{(n)}$ denote the coordinates of the vector of order statistics; we assume the existence of a density $f^{(n)}(t_1, ..., t_n)$, so that (2.8) holds; under this assumption, it must be necessarily $P\{T_{(1)} < ... < T_{(n)}\} = 1$: simultaneous failures happen with null probability.

Consider then a duration experiment where the units $U_1, ..., U_n$ are new at time 0 and are simultaneously put under test, progressively recording all the subsequent failure times. At any time $t > 0$ the available statistical observation is a history of the form (2.37).

Definition 2.29. For $h = 1, ..., n - 1$, the *multivariate conditional hazard rate* of a unit U_j, surviving at t, given the history in (2.37), is defined as the limit

$$\lambda_t^{(n-h)}(t_1, ..., t_h) \equiv$$

$$\lim_{\delta \to 0+} \frac{1}{\delta} P\{T_{j_l} < t + \delta | T_{i_1} = t_1, ..., T_{i_h} = t_h, T_{j_1} > t, ..., T_{j_{n-h}} > t\}.$$

where $l = 1, 2, ..., n - h$. For $h = 0$, we define

$$\lambda_t^{(n)} \equiv \lim_{\delta \to 0+} \frac{1}{\delta} P\{T_j < t + \delta | T_1 > t, ..., T_n > t\}.$$

The suffix $(n - h)$ denotes the number of the units which are still surviving at time t.

Note that, due to exchangeability, the above limit does not depend on the choice of the index $j \in \{j_1, ..., j_{n-h}\}$. In view of this assumption, moreover, there is no practical interest in recording also the "identities" of the units which fail. Thus we can consider that, in place of (2.37), the observed history is one of the form

$$\mathfrak{h}_t \equiv \{T_{(1)} = t_1, ..., T_{(h)} = t_h, T_{(h+1)} > t\} \tag{2.40}$$

where $0 < t_1 \leq ... \leq t_h$.

Also in the definition of the multivariate conditional hazard rate functions we do not need to refer to the information about which units are still surviving at any time t. Indeed, since we assumed $P\{T_{(1)} < ... < T_{(n)}\} = 1$, we can equivalently define $\lambda_t^{(n-h)}(t_1, ..., t_h)$ $(h = 1, ..., n - 1)$ and $\lambda_t^{(n)}$ in terms of the vector of the order statistics by letting

$$\lambda_t^{(n-h)}(t_1, ..., t_h) =$$

$$\lim_{\delta \to 0^+} \frac{1}{\delta(n - h)} P\{T_{(h+1)} \leq t + \delta | T_{(1)} = t_1, ..., T_{(h)} = t_h, T_{(h+1)} > t\} \tag{2.41}$$

and, for $h = 0$,

$$\lambda_t^{(n)} = \lim_{\delta \to 0^+} \frac{1}{n\delta} P\{T_{(1)} \leq t + \delta | T_{(1)} > t\}. \tag{2.42}$$

In the present cases, the above limits, defining $\lambda_t^{(n-h)}(t_1, ..., t_h)$ and $\lambda_t^{(n)}$, exist and expressions for them are provided by the following result.

Proposition 2.30. *For* $1 \leq h \leq n - 1$, $0 < t_1 \leq ... \leq t_h \leq t$,

$$\lambda_t^{(n-h)}(t_1, ..., t_h) =$$

$$\frac{\int_t^\infty \cdots \int_t^\infty f^{(n)}(t_1, ..., t_h, t, t_{h+2}, ..., t_n) \, dt_{h+2}...dt_n}{\int_t^\infty \cdots \int_t^\infty f^{(n)}(t_1, ..., t_h, t_{h+1}, t_{h+2}, ..., t_n) \, dt_{h+1}dt_{h+2}...dt_n} \tag{2.43}$$

Furthermore

$$\lambda_t^{(n)} = \frac{\int_t^\infty \cdots \int_t^\infty f^{(n)}(t, t_2, ..., t_n) \, dt_2...dt_n}{\int_t^\infty \cdots \int_t^\infty f^{(n)}(t_1, t_2, ..., t_n) \, dt_1 dt_2...dt_n} \tag{2.44}$$

Proof. For $1 \leq h \leq n - 1$, $0 < t_1 \leq ... \leq t_h \leq t$, consider the conditional survival function of the residual life-times of the $(n - h)$ units which are still surviving at t, given an observed history of the form (2.37) or (2.40).

By specializing (2.20), we obtain that, as a function of the ages $s_{h+1}, ..., s_n$, this is given by

$$P\{T_{j_1} > s_{h+1}, ..., T_{j_{n-h}} > s_n | \mathfrak{h}_t\} =$$

$$\frac{\int_{s_{h+1}}^{\infty} \cdots \int_{s_n}^{\infty} f^{(n)}(t_1, ..., t_h, t_{h+1}, ..., t_n) dt_{h+1} ... dt_n}{\int_t^{\infty} \cdots \int_t^{\infty} f^{(n)}(t_1, ..., t_h, t_{h+1}, ..., t_n) dt_{h+1} ... dt_n}$$

In order to obtain the corresponding one-dimensional marginal we then have to compute the latter in the vector of ages $(s, t, ..., t)$, which yields

$$\frac{\int_s^{\infty} \int_t^{\infty} \cdots \int_t^{\infty} f^{(n)}(t_1, ..., t_h, t_{h+1}, ..., t_n) dt_{h+1} ... dt_n}{\int_t^{\infty} \cdots \int_t^{\infty} f^{(n)}(t_1, ..., t_h, t_{h+1}, ..., t_n) dt_{h+1} ... dt_n}$$

By combining the latter with Definition 2.29, we get

$$\lambda_t^{(n-h)}(t_1, ..., t_h) =$$

$$\lim_{\delta \to 0+} \frac{1}{\delta} \frac{\int_t^{t+\delta} \int_t^{\infty} \cdots \int_t^{\infty} f^{(n)}(t_1, ..., t_h, t_{h+1}, ..., t_n) dt_{h+1} ... dt_n}{\int_t^{\infty} \cdots \int_t^{\infty} f^{(n)}(t_1, ..., t_h, t_{h+1}, ..., t_n) dt_{h+1} ... dt_n}$$

whence we obtain (2.43). For $h = 0$, the expression (2.44) is obtained similarly.

$\square$

Remark 2.31. In the case when $T_1, ..., T_n$ are i.i.d. with a one-dimensional hazard rate function $r(t)$, Equations (2.43) and (2.44)) trivially reduce to

$$\lambda_t^{(n-h)}(t_1, ..., t_h) = r(t); \lambda_t^{(n)} = r(t) \tag{2.45}$$

Remark 2.32. Equation (2.42) shows that $n\lambda_t^{(n)}$ is the one-dimensional failure rate function of the variable

$$T_{(1)} \equiv \min_{1 \leq j \leq n} T_j.$$

So we can also read (2.42) as

$$\lambda_t^{(n)} = \frac{f_{T_{(1)}}(t)}{nP\{T_{(1)} > t\}} \text{ or } \lambda_t^{(n)} = \frac{f_{T_{(1)}}(t)}{n \exp\{-n \int_0^t \lambda_u^{(n)} du\}} \tag{2.46}$$

In an analogous way, we can read (2.41) in the form

$$\lambda_t^{(n-h)}(t_1, ..., t_h) =$$

$$\frac{f_{T_{(h+1)}}(t | T_{(1)} = t_1, ..., T_{(h)} = t_h, T_{(h+1)} > t_h)}{(n-h)P\{T_{(h+1)} > t | T_{(1)} = t_1, ..., T_{(h)} = t_h, T_{(h+1)} > t_h\}}. \tag{2.47}$$

We now proceed to present a few examples.

Some of the examples show the special form taken by the m.c.h.r. functions in some remarkable special cases; other examples, alternatively, show how some models of interest can be defined, in a natural way, starting from the assignment of the set of the m.c.h.r. functions.

Example 2.33. Consider two units undergoing a change in the level of stress at a random time σ as in Example 2.19. From Definition 2.29 we can obtain

$$\lambda_t^{(1)}(t_1) = P\{\sigma > t | T_1 = t_1, T_2 > t\}r_1(t) + P\{\sigma \leq t | T_1 = t_1, T_2 > t\}r_2(t)$$

$$\lambda_t^{(2)} = P\{\sigma > t | T_1 > t, T_2 > t\}r_1(t) + P\{\sigma \leq t | T_1 > t, T_2 > t\}r_2(t)$$

(see also Exercise 2.71).

Example 2.34. In the case of two individuals in Example 2.20, where

$$\overline{F}^{(2)}(s_1, s_2) = p_0(0)\overline{F}_1(s_1)\overline{F}_1(s_2)+$$

$$\frac{1}{2}p_0(1)\left\{\overline{F}_0(s_1)\overline{F}_1(s_2) + \overline{F}_0(s_2)\overline{F}_1(s_1)\right\} + p_0(2)\overline{F}_0(s_1)\overline{F}_0(s_2),$$

we have

$$\lambda_t^{(2)} = r_1(t)P\{M = 0 | T_1 > t, T_2 > t\}$$

$$+\frac{1}{2}\{r_1(t) + r_0(t)\}P\{M = 1 | T_1 > t, T_2 > t\}+$$

$$r_0(t)P\{M = 2 | T_1 > t, T_2 > t\}$$

Example 2.35. Consider the situation of a Schur-constant joint density already analyzed in Example 1.52 of Chapter 1. Let then $T_1, ..., T_n$ have a joint density such that, for a suitable function $\phi_n : R_+ \to R_+$,

$$f^{(n)}(t_1, ..., t_n) = \phi_n(\sum_{i=1}^{n} t_i).$$

By applying the formula (2.9), we can easily check that this is equivalent to

$$\overline{F}^{(n)}(s_1, ..., s_n) = \Phi(\sum_{i=1}^{n} s_i)$$

with $\overline{\Phi} : \mathbb{R}_+ \to [0,1]$ such that $\phi_n(t) = (-1)^n \frac{d^n}{ds^n}\overline{\Phi}|_{s=t}$. Moreover, by (2.41), we can find (see Caramellino and Spizzichino, 1996)

$$\lambda_t^{(n-h)}(t_1,...,t_h) = -\frac{\frac{d^{h+1}}{dy^{h+1}}\overline{\Phi}(y)}{\frac{d^h}{dy^h}\overline{\Phi}(y)}$$

with

$$y = \sum_{i=1}^{h} t_i + (n-h)t.$$

Example 2.36. Think of two similar individuals, between whom there is a strong rivalry, so that each individual undergoes a stress which is proportional to the total age cumulated by the other individual. We model this by assuming that their lifetimes T_1 and T_2 are exchangeable random quantities with a joint distribution characterized by m.c.h.r's of the form

$$\lambda_t^{(2)}(\theta) = \theta t, \lambda_t^{(1)}(t_1|\theta) = \theta t_1$$

where the "baseline" hazard $\theta > 0$ is a known constant. For any fixed $t > 0$ and $\theta > 0$, $\lambda_t^{(1)}(t_1|\theta)$ is an increasing function of t_1 in the interval $[0,t)$ and this can be looked at as a form of negative dependence between T_1 and T_2.

Example 2.37. An important (and often analyzed) class of multivariate reliability models arises from the assumption that the instantaneous conditional failure rate of a component surviving at t does not depend on the failure-times of previously failed components: the failure rate is assumed to depend only on t, on the number, and on the identities of all the components still surviving at t ("working set"). Such a case can be seen as a generalization of models considered in Kopocinska and Kopocinski (1980); Ross (1984); Shechner (1984) and references cited therein (special cases of this are the "load-sharing models" and the "Ross models").

In order to get, for such a model, a case of exchangeability, we must impose that all the components surviving at t have a same failure rate and that this failure rate depends only on the total number $(n-h)$, but not on the "identities", of the surviving components. Summarizing, we consider those exchangeable models arising from the position

$$\lambda_t^{(n-h)}(t_1,...,t_h) = \lambda_t^{(n-h)}, \ h = 1,...,n-1. \tag{2.48}$$

As special cases of (2.48) one has the "linear-breakdown" models, where $\lambda_t^{(n-h)} = \frac{L(t)}{n-h}$ for a suitable function $L(t)$. By imposing, instead, time-homogeneity we get exchangeable Ross models, where $\lambda_t^{(n-h)}$ is independent of t. By combining the latter two conditions we obtain the well-known special case

$$\lambda_t^{(n-h)} = \frac{\theta}{n-h},$$

where θ is a given positive quantity.

Remark 2.38. Note that the condition (2.48) does not mean that $T_1, ..., T_n$ are independent. Indeed $\lambda_t^{(n-h)}(t_1, ..., t_h)$ is a function of h (beside being in general a function of t), while, as remarked above a necessary and sufficient condition for stochastic independence is that $\lambda_t^{(n-h)}$ does not depend on h.

In the particular cases of parametric models that were considered in the last section, $\lambda_t^{(n-h)}(t_1, ..., t_h)$ and $\lambda_t^{(n)}$ can be given a particularly significant form:

Proposition 2.39. *Let the joint density function be of the form (2.22). Then it is*

$$\lambda_t^{(n-h)}(t_1, ..., t_h) = \int_L r(t|\theta)\pi(\theta|\mathfrak{h}_t)d\theta \qquad (2.49)$$

$$\lambda_t^{(n)} = \int_L r(t|\theta)\pi(\theta|T_{(h+1)} > t)d\theta \qquad (2.50)$$

where $\mathfrak{h}_t$ is as (2.40) and where $\pi(\theta|\mathfrak{h}_t)$ and $\pi(\theta|T_{(h+1)} > t)$ can be obtained as special cases of (2.24).

Proof. Replace $f^{(n)}$ in (2.43) with the r.h.s. of (2.22) and change the order of integration. This yields

$$\lambda_t^{(n-h)}(t_1, ..., t_h) =$$

$$\frac{\int_L \left[\int_t^\infty \cdots \int_t^\infty g(t_1|\theta)...g(t_h|\theta)g(t|\theta)g(t_{h+2}|\theta)...g(t_n|\theta)dt_{h+2}...dt_n\right] d\theta}{\int_L \left[\int_t^\infty \cdots \int_t^\infty g(t_1|\theta)...g(t_h|\theta)g(t_{h+1}|\theta)g(t_{h+2}|\theta)...g(t_n|\theta)dt_{h+1}dt_{h+2}...dt_n\right] d\theta} =$$

$$\frac{\int_L \left[g(t_1|\theta)...g(t_h|\theta)\frac{g(t|\theta)}{\overline{G}(t|\theta)}\overline{G}(t|\theta)\,\overline{G}(t_{h+2}|\theta)...\overline{G}(t_n|\theta)\right] d\theta}{\int_L \left[g(t_1|\theta)...g(t_h|\theta)\overline{G}(t|\theta)\,\overline{G}(t_{h+2}|\theta)...\overline{G}(t_n|\theta)\right] d\theta} =$$

$$\int_L r(t|\theta)\frac{g(t_1|\theta)...g(t_h|\theta)\overline{G}(t|\theta)\,\overline{G}(t_{h+2}|\theta)...\overline{G}(t_n|\theta)}{\int_L g(t_1|\theta)...g(t_h|\theta)\overline{G}(t|\theta)\,\overline{G}(t_{h+2}|\theta)...\overline{G}(t_n|\theta)}d\theta.$$

By adapting (2.24) to the special case when $D = \mathfrak{h}_t$ as in (2.40), we see that the latter expression becomes

$$\lambda_t^{(n-h)}(t_1, ..., t_h) = \int_L r(t|\theta)\pi(\theta|\mathfrak{h}_t)d\theta;$$

A similar argument holds for (2.50). $\qquad\qquad\square$

Note that Proposition 2.39 could also be directly proved by making use of Proposition 2.15.

Now contrast (2.49) and (2.45); the heuristic interpretation of (2.49) is rather immediate: if we knew the value θ taken by the parameter Θ, then, due to conditional independence given Θ, the value of the hazard rate for a unit surviving at t would be equal to $r(t|\theta)$; when Θ is unknown the observed history (2.40) is to be taken into account to update the distribution of Θ; in this respect, see also Remark 2.21.

Example 2.40. (Proportional hazard models). This is the case when $r(t|\theta)$ in (2.49) is of the form $r(t|\theta) = \theta\rho(t)$; whence we have

$$\lambda_t^{(n-h)}(t_1, ..., t_h) = \rho(t)\int_L \theta\pi(\theta|\mathfrak{h}_t)d\theta = \rho(t)\mathbb{E}(\Theta|\mathfrak{h}_t) \qquad (2.51)$$

Remark 2.41. The condition (2.22), for which the m.c.h.r. functions have been derived just above, can give rise to relevant cases of positive dependence. This will be rendered in more precise terms in Chapter 3 (Subsection 3.3).

It is easy to understand that cases of positive dependence are also those described in the above Example 2.37, when the function $\lambda_t^{(n-h)}$ is increasing in h.

However one can intuitively see that the former and the latter cases are different in nature: in the former case, dependence is due to "increase of information," that is to conditional independence given the same unknown quantity; in the latter case positive dependence is just due to a "physical" interaction among components.

Such a difference will have an impact on the analysis of multivariate aging.

Before continuing it is convenient to state the following Lemma, which can be simply proved by applying the "product rule" formula of conditional probabilities and by taking into account (2.41) and (2.42).

Lemma 2.42. *For $h = 1, ..., n - 1, 0 < t_1 < ... < t_h < t$, one has*

$$\lim_{\delta\to 0} \frac{P\{t_1 \leq T_{(1)} < t_1 + \delta, ..., t_h \leq T_{(h)} < t_h + \delta, T_{(h+1)} > t\}}{\delta^h}$$

$$= n!\,\lambda_{t_1}^{(n)}\prod_{j=2}^{h}\lambda_{t_j}^{(n-j+1)}(t_1, ..., t_{j-1})\times$$

$$\exp\left\{-n\int_0^{t_1}\lambda_u^{(n)}du - \sum_{j=2}^{h}(n-j-1)\int_{t_{j-1}}^{t_j}\lambda_u^{(n-j+1)}(t_1, ..., t_{j-1})du\right\}$$

Proposition 2.39 is a special case of the following more general situation, which is also of potential interest in several applications: $T_1, ..., T_n$ are conditionally exchangeable, given the value θ taken by a parameter Θ ($\theta \in L \subset \mathbb{R}^d$ for some $d = 1, 2, ...$), with a joint distribution characterized by c.m.h.r.'s $\lambda_t^{(n)}(\theta)$ and $\lambda_t^{(n-h)}(\cdot|\theta)$ depending on θ.

In such cases it is of interest to obtain the ("predictive") m.c.h.r. functions $\lambda_t^{(n)}$ and $\lambda_t^{(n-h)}(\cdot)$ of $T_1, ..., T_n$ in terms of the "initial" density of Θ and of the m.c.h.r.'s $\lambda_t^{(n)}(\theta), \lambda_t^{(n-h)}(\cdot|\theta)$ conditional on Θ ($h = 1, 2, ..., n - 1$).

This problem is in general solved by the following

Theorem 2.43. *For $h = 1, ..., n - 1$, one has*

$$\lambda_t^{(n-h)}(t_1, ..., t_h) = \int_L \lambda_t^{(n-h)}(t_1, ..., t_h|\theta)\pi(\theta|\mathfrak{h}_t)d\theta \qquad (2.52)$$

and

$$\lambda_t^{(n)} = \int_L \lambda_t^{(n)}(\theta)\pi(\theta|T_{(1)} > t)d\theta. \qquad (2.53)$$

Proof. We start by proving (2.53). By Bayes' theorem one has

$$\pi(\theta|T_{(1)} > t) = \frac{\pi_0(\theta)\,\overline{F}_{T_{(1)}}(t|\theta)}{\int_L \pi_0(\theta)\,\overline{F}_{T_{(1)}}(t|\theta)\,d\theta}.$$

Then, in order to prove (2.53), we have to check the identity

$$\lambda_t^{(n)} = \frac{\int_L \lambda_t^{(n)}(\theta)\overline{F}_{T_{(1)}}(t|\theta)\,\pi_0(\theta)\,d\theta}{\int_L \overline{F}_{T_{(1)}}(t|\theta)\,\pi_0(\theta)\,d\theta}. \qquad (2.54)$$

On the other hand, as a consequence of Equation (2.42), we can write

$$\lambda_t^{(n)}(\theta) = \frac{1}{n}\frac{f_{T_{(1)}}(t|\theta)}{\overline{F}_{T_{(1)}}(t|\theta)}$$

whence (2.54) becomes

$$\lambda_t^{(n)} = \frac{1}{n}\frac{\int_L f_{T_{(1)}}(t|\theta)\pi_0(\theta)\,d\theta}{\int_L \overline{F}_{T_{(1)}}(t|\theta)\,\pi_0(\theta)\,d\theta} = \frac{1}{n}\frac{f_{T_{(1)}}(t)}{\overline{F}_{T_{(1)}}(t|\theta)}$$

which is seen to be true by applying Equation (2.42) to the unconditional distribution of $T_{(1)}$.

Equation (2.52) can be proved in a similar way, as follows.

First we need to extend Equation (2.24) to the present more general case.

Denoting by $f^{(n)}(t_1, ..., t_n|\theta)$ the conditional joint density of $T_1, ..., T_n$, given $\Theta = \theta$ (with $\theta \in L$), we first note that, for $\mathfrak{h}_t$ as in (2.40)

$$\pi(\theta|\mathfrak{h}_t) = \frac{\int_t^\infty \cdots \int_t^\infty f^{(n)}(t_1, .., t_h, t_{h+1}, .., t_n|\theta)\pi_0(\theta)\, dt_{h+1}...dt_n\, d\theta}{\int_L \int_t^\infty \cdots \int_t^\infty f^{(n)}(t_1, .., t_h, t_{h+1}, .., t_n|\theta)\pi_0(\theta)\, dt_{h+1}...dt_n\, d\theta}.$$

This can be obtained by Bayes theorem taking into account that, in view of Lemma 2.42, the likelihood associated with the result $\mathfrak{h}_t$ indeed has the form

$$\ell(\theta) = \int_t^\infty \cdots \int_t^\infty f^{(n)}(t_1, ..., t_h, u_{h+1}, ..., u_n|\theta)du_{h+1}...du_n.$$

Thus our task becomes one of checking that

$$\lambda_t^{(n-h)}(t_1, ..., t_h) =$$

$$\frac{\int_L \lambda_t^{(n-h)}(t_1, .., t_h|\theta)\int_t^\infty \cdots \int_t^\infty f^{(n)}(t_1, .., t_h, t_{h+1}, u|\theta)\pi_0(\theta)\, dt_{h+1}du\, d\theta}{\int_L \int_t^\infty \cdots \int_t^\infty f^{(n)}(t_1, .., t_h, t_{h+1}, u|\theta)\pi_0(\theta)\, dt_{h+1}du\, d\theta} \quad (2.55)$$

where, for brevity, in the notation we replaced $u_{h+2}, ..., u_n$ by u and $du_{h+2}...du_n$ by du. From Equation (2.43), we get

$$\lambda_t^{(n-h)}(t_1, ..., t_h|\theta)\int_t^\infty \cdots \int_t^\infty f^{(n)}(t_1, ..., t_h, t_{h+1}, u|\theta)dt_{h+1}du =$$

$$(n-h)\int_t^\infty \cdots \int_t^\infty f^{(n)}(t_1, ..., t_h, t, u|\theta)du.$$

Then the r.h.s. of (2.55) becomes

$$(n-h)\frac{\int_L[\int_t^\infty \cdots \int_t^\infty f^{(n)}(t_1, ..., t_h, t, u|\theta)du]\pi_0(\theta)\, d\theta}{\int_L[\int_t^\infty \cdots \int_t^\infty f^{(n)}(t_1, ..., t_h, t_{h+1}, u|\theta)dt_{h+1}du]\pi_0(\theta)\, d\theta} =$$

$$(n-h)\frac{\int_t^\infty \cdots \int_t^\infty[\int_L f^{(n)}(t_1, ..., t_h, t, u|\theta)\pi_0(\theta)\, d\theta]du}{\int_t^\infty \cdots \int_t^\infty[\int_L f^{(n)}(t_1, ..., t_h, t_{h+1}, u|\theta)\pi_0(\theta)\, d\theta]dt_{h+1}du} =$$

$$(n-h)\frac{\int_t^\infty \cdots \int_t^\infty f^{(n)}(t_1, ..., t_h, t, u)du}{\int_t^\infty \cdots \int_t^\infty f^{(n)}(t_1, ..., t_h, t, u)dt_{h+1}du},$$

since

$$f^{(n)}(t_1, ..., t_n) = \int_L f^{(n)}(t_1, ..., t_n|\theta)\pi_0(\theta)\, d\theta.$$

Then (2.55) is proved by taking again into account Equation (2.43). $\qquad\square$

Remark 2.44. Let π_0 be the initial density of Θ and suppose the history $\mathfrak{h}_t$ as in (2.40) has been observed ($h = 0, 1, ..., n - 1$). By Lemma 2.42 we see that the posterior density of Θ can be specified by

$$\pi(\theta|\mathfrak{h}_t) \propto \lambda_{t_1}^{(n)}(\theta) \prod_{j=2}^{h} \lambda_{t_h}^{(n-j+1)}(\theta|t_1, ..., t_{j-1}) \times$$

$$\exp\{-n \int_0^{t_1} \lambda_u^{(n)}(\theta)du - \sum_{j=2}^{h}(n - j - 1) \int_{t_{j-1}}^{t_j} \lambda_u^{(nj+1)}(\theta|t_1, ..., t_{j-1})du$$

$$-(n - h) \int_{t_h}^{t} \lambda_u^{(n-h)}(\theta|t_1, ..., t_h)du\} \tag{2.56}$$

Note that Equation (2.52) provides an appropriate generalization of (2.49).

Remark 2.45. As mentioned, it is often the case that Θ has to be replaced by a stochastic process Θ_t. Indeed Θ, or some of its coordinates, can be thought of as describing environmental conditions which may be liable to undergo time-variations as also mentioned at the end of the previous section. In such cases the equations in (2.49) and (2.50) have to be suitably modified to more general formulae.

In order to obtain such formulae a setting of stochastic processes and appropriate general assumptions are needed. However it is still possible to see directly, by means of a heuristic reasoning, what in tractable cases the appropriate generalization of (2.49) and (2.50) should be: the conditional distribution of Θ, given the observed history (2.40) is to be replaced by the conditional distribution of Θ_t, given the observed history (2.40). Here we shall not consider such a generalization.

Rigorous results can be obtained as particular cases of a general theorem from the theory of point processes (see e.g. Bremaud (1981); the techniques of stochastic filtering just serve to obtain the conditional distribution of Θ_t, given the observed history; for a deeper mathematical discussion on this topic see Arjas (1989).

Remark 2.46. Often it is natural to conceive the units (or "individuals"), whose lifetimes $T_1, ..., T_n$ we are studying, as "parts" or "components" of a single system. This is the case, for instance, in Example 2.37. It can be true for biological applications as well, where it is impossible to abstract individuals of a population from their common environment, and in which also interactions among different members are necessarily present. We remark however that a specific feature of our treatment is that, even when the individuals are part of the same system, we are not interested in predicting the evolution of the system as such (e.g. to estimate its own failure-time); we are interested in the lifetimes of any single unit. This means that we completely ignore the structure function of the system.

Remark 2.47. In modeling dependence among lifetimes, we are rather interested in studying the effects both of the common environment and of the stress (or the help) that affects any single unit at a generic time-instant t as the result of the behavior of all the other units in the interval $[0, t)$. It is to be noted that dependence due to informational effects about the common environment and dependence due to physical effects of interactions among units may overlap in determining the kind of stochastic dependence that is described by means of the m.c.h.r. functions. For the case when environmental factors are constant in time, this is better explained by means of Theorem 2.43.

2.4   Further aspects of m.c.h.r.

In this section we focus attention on some further aspects of interest related to the notion of multivariate conditional hazard rate. For this purpose, we keep the same setting and same notation as in Section 2.3, where a longitudinal observation of failure data was considered: individuals $U_1, ..., U_n$ are units which start living at a same epoch 0; the quantities $T_1, ..., T_n$ denote the lifetimes of $U_1, ..., U_n$ and up to any time t a dynamic history of the form $\mathfrak{h}_t \equiv \{T_{(1)} = t_1, ..., T_{(h)} = t_h, T_{(h+1)} > t\}$ as in (2.40) is observed: $T_{(1)}, ..., T_{(n)}$ are the order statistics, h is the number of failures observed up to $t, t_1 \leq ... \leq t_h$ are the corresponding failure times and all the units surviving at t share the same age t. h can be seen as the observed value, at time t, of the stochastic process H_t defined as follows

$$H_t \equiv \sum_{j=1}^{n} \mathbf{1}_{\{T_j \leq t\}}, \forall t \geq 0 \tag{2.57}$$

where the symbol $\mathbf{1}_A$ denotes the indicator of the event A.

H_t counts the number of failures observed up to time t; of course $H_0 = 0$ and the trajectories of H_t are stepwise constant, non-decreasing, functions on $[0, +\infty)$, with jumps at the instants $T_{(1)} \leq ... \leq T_{(n)}$.

2.4.1   On the use of the m.c.h.r. functions

We saw, in Proposition 2.30, how $\lambda_t^{(n-h)}(t_1, ..., t_h)$ $(0 \leq h \leq n - 1; 0 < t_1 \leq ... \leq t_h)$ are to be computed in terms of the joint density function $f^{(n)}$.

Conversely, we can see how $f^{(n)}$ can be computed starting from the knowledge of the m.c.h.r. functions. Indeed the joint density function is determined by the family of functions $\{\lambda_t^{(k)}\}_{k=1,...,n;t\geq0}$ and we can in particular obtain the following Proposition, as a direct consequence of Lemma 2.42.

Proposition 2.48. *If the limits defining $\lambda_t^{(n)}$ and $\lambda_t^{(n-h)}(\cdot)$ $(h = 1, 2, ..., n-1)$ exist, the joint distribution of $T_1, ..., T_n$ admits a joint density $f^{(n)}$. Denoting by $(t_{(1)}, ..., t_{(n)})$ the vector of order statistics of $(t_1, ..., t_n)$, the following relation holds:*

$$f^{(n)}(t_1, ..., t_n) = \lambda_{t_{(1)}}^{(n)} \prod_{h=2}^{n} \lambda_{t_{(h)}}^{(n-h+1)}(t_{(1)}, ..., t_{(h-1)}) \times \exp\left\{-n \int_0^{t_{(1)}} \lambda_u^{(n)} du\right\} \times$$

$$\exp\left\{-\sum_{h=2}^{n}[n-(h-1)] \int_{t_{(h-1)}}^{t_{(h)}} \lambda_u^{(n-h+1)}(t_{(1)}, ..., t_{(h-1)}) du\right\} \qquad (2.58)$$

Note that Equation (2.58) is a multivariate analogue of (2.3).

Example 2.49. Let $\theta > 0$ be a given number and consider the case

$$\lambda_t^{(n)} = \theta, t \geq 0$$

$$\lambda_t^{(n-h)}(t_1, ..., t_h) = \theta, 0 \leq h \leq n-1, 0 < t_1 \leq ... \leq t_h \leq t.$$

Then (2.58) yields

$$f^{(n)}(t_1, ..., t_n) = f^{(n)}(t_{(1)}, ..., t_{(n)}) = \theta^n \exp\{-\theta \cdot \sum_{j=1}^{n} t_j\}.$$

i.e. $T_1, ..., T_n$ are i.i.d. with exponential distribution of parameter θ.

More generally, (2.58) shows that the case of i.i.d. lifetimes is characterized by m.c.h.r.'s of the form in (2.45).

Example 2.50. For the case $\lambda_t^{(n-h)}(t_1, ..., t_h) = \frac{\theta}{n-h}$ (linear breakdown model), considered in Example 2.37 we can obtain, from (2.58), that the corresponding joint density is

$$f^{(n)}(t_1, ..., t_n) = \frac{\theta^n}{n!} \exp\{-\theta \cdot t_{(n)}\}$$

The case considered just above (as well as the case in Example 2.36, for instance) provides an example of the following fact: in many situations, which arise in the applications of reliability, dependence among lifetimes can be modeled in a completely natural way in terms of the m.c.h.r. functions. In the same cases, on the contrary, it can take a while to figure out what the corresponding joint density or the joint survival function should be.

This fact allows us to focus one reason of interest in Proposition 2.48, which actually permits one to obtain the joint density starting from the m.c.h.r. functions.

We also notice that, among the different tools which characterize an absolutely continuous probability distribution on $\mathbb{R}_+^n$, the set of m.c.h.r. functions is, in a sense, the most natural way to describe the conditional distribution of residual lifetimes $T_{j_1} - t, ..., T_{j_{n-h}} - t$, given an observed history of the type (2.40).

Indeed, for $h = 0, 1, ..., n-1, 0 < t_1 \leq ... \leq t_h \leq t$, let us consider the m.c.h.r. functions corresponding to such a conditional distribution, i.e., for $r = 1, ..., n - h + 1, 0 < u_1 \leq ... \leq u_r \leq s$, let us consider

$$\widetilde{\lambda}_s^{((n-h)-r)}(u_1, ..., u_r | h; t_1, ..., t_h; t) \equiv$$

$$\lim_{\delta \to 0} \frac{1}{\delta} P\{T_{(h+r+1)} - t \leq s + \delta | T_{(1)} = t_1, ..., T_{(h)} = t_h;$$

$$T_{(h+1)} - t = u_1, ..., T_{(h+r)} - t = u_r, T_{(h+r+1)} - t > s\}$$

and, for the case $r = 0$,

$$\widetilde{\lambda}_s^{(n-h)}(|h; t_1, ..., t_h; t) \equiv$$

$$\lim_{\delta \to 0} \frac{1}{\delta} P\{T_{(h+1)} - t \leq s + \delta | T_{(1)} = t_1, ..., T_{(h)} = t_h; T_{(h+1)} - t > s\}$$

By the very definition of m.c.h.r. function, we easily see that

$$\widetilde{\lambda}_s^{(n-h)}(|h; t_1, ..., t_h; t) = \lambda_{s+t}^{(n-h)}(t_1, ..., t_h)$$

and, for $r = 1, ..., n - h + 1$,

$$\widetilde{\lambda}_s^{((n-h)-r)}(u_1, ..., u_r | h; t_1, ..., t_h; t) = \lambda_{t+s}^{(n-(h+r))}(t_1, ..., t_r, u_1 + t, ..., u_r + t).$$
$$(2.59)$$

Note that Equation (2.59) can be seen as the direct multivariate analogue of the formula $r_t(s) = r(t + s)$ which holds for one-dimensional hazard rate functions, where r is the failure rate function of a lifetime T and $r_t(s)$ denotes the conditional failure rate of a residual lifetime $T - t$, conditional on $\{T > t\}$.

On the contrary, the set of m.c.h.r. functions is not a suitable tool for deriving marginal distributions or conditional distributions of a subset of lifetimes given an observation of the form $\{T_{i_1} = t_1, ..., T_{i_h} = t_h\}$ for the other lifetimes.

Indeed the multivariate conditional hazard at any time $t > 0$, only allows predictions about the behavior, just after t, of those units which still survive at t, while it does not allow probabilistic assessment about events before t.

One aspect of the above is that, in particular, no simple equation relates m.c.h.r. functions of m-dimensional marginal distributions to the m.c.h.r. functions of the original n-dimensional distribution $(m < n)$.

Example 2.51. Consider again the linear breakdown Ross model with $\theta > 0$ given and $n = 3$. In such a case

$$\lambda_t^{(1)}(t_1, t_2) = \theta, \qquad \lambda_t^{(2)}(t_1) = \frac{\theta}{2}, \qquad \lambda_t^{(3)} = \frac{\theta}{3},$$

$$f^{(3)}(t_1, t_2, t_3) = \frac{\theta^3}{3!} \exp\{-\theta \cdot t_{(3)}\}.$$

As far as the two-dimensional marginal density function is concerned, we have

$$f^{(2)}(t_1, t_2) = \int_0^\infty f^{(3)}(t_1, t_2, t_3) dt_3$$

$$= \int_0^{t_1 \vee t_2} f^{(3)}(t_1, t_2, t_3) dt_3 + \int_{t_1 \vee t_2}^\infty f^{(3)}(t_1, t_2, t_3) dt_3$$

$$= \int_0^{t_1 \vee t_2} \frac{\theta^3}{3!} \exp\{-\theta \cdot t_1 \vee t_2\} dt_3 + \int_{t_1 \vee t_2}^\infty \frac{\theta^3}{3!} \exp\{-\theta \cdot t_3\} dt_3$$

$$= \frac{\theta^2}{3!} \exp\{-\theta \cdot t_1 \vee t_2\}(\theta \cdot t_1 \vee t_2 + 1).$$

The m.c.h.r. functions of the pair (T_1, T_2) can be obtained by applying Equations (2.41) and (2.42) to the above two-dimensional joint density. Some lengthy computations give

$$\lambda_t^{(1)}(t_1) = \theta \frac{(t \cdot \theta + 1)}{(t \cdot \theta + 2)}, \quad \lambda_t^{(2)} = \frac{\theta}{2} \frac{(t \cdot \theta + 2)}{(t \cdot \theta + 3)}. \tag{2.60}$$

This still defines a non-homogeneous exchangeable model where the m.c.h.r. do not depend on past failure times.

2.4.2 Dynamic histories, total time on test statistic and total hazard transform

In the statistical analysis of a dynamic history the following definitions are of interest.

Definition 2.52. The total time on test (in short TTT) process is the stochastic process defined by

$$Y_t = \sum_{i=1}^n \left(T_{(i)} \wedge t \right), t \geq 0 \tag{2.61}$$

where $a \wedge b$ denotes the minimum between a and b, as usual.

In terms of the process H_t, we can give Y_t the following equivalent expressions:

$$Y_t = \sum_{i=1}^{H_t} T_{(i)} + (n - H_t) \cdot t; \quad Y_t = \int_0^t (n - H_s)\, ds \qquad (2.62)$$

Note that Y_t quantifies the total amount of life spent by individuals $U_1, ..., U_n$ in the time interval $[0, t]$.

For $h = 1, 2, ..., n$, consider the random variable

$$Y_{(h)} \equiv Y_{T_{(h)}}$$

$Y_{(h)}$ is the total time on test cumulated by $U_1, ..., U_n$ until the h-th failure, i.e. in the random time interval $[0, T_{(h)}]$.

We can also write

$$Y_{(h)} = \sum_{i=1}^{h} T_{(i)} + (n - h) \cdot T_{(h)} \qquad (2.63)$$

Definition 2.53. For $h = 1, 2, ..., n$, the h-th normalized spacing between order statistics is the random variable defined by

$$C_h \equiv (n - h + 1) \cdot \left(T_{(h)} - T_{(h-1)}\right) \qquad (2.64)$$

where $T_{(0)} \equiv 0$.

C_h is the total time on test cumulated by surviving individuals between the $(h-1)$-th and h-th failure; from (2.63) and Definition 2.53, we immediately see that $Y_{(h)}$ can be decomposed in terms of the normalized spacings as follows

$$Y_{(h)} = \sum_{i=1}^{h} C_i. \qquad (2.65)$$

Example 2.54. In a life testing experiment on 6 similar units, the following failure data have been observed:

$t_{(1)} = 9,000\ h$, $t_{(2)} = 16,000\ h$, $t_{(3)} = 20,000\ h$, $t_{(4)} = 24,600\ h$, $t_{(5)} = 28,000\ h$, $t_{(6)} = 30,400\ h$.

This result yields

$C_1 = 54,000$; $C_2 = 35,000$; $C_3 = 16,000$; $C_4 = 13,800$; $C_5 = 7,800$; $C_6 = 2,400$.

$Y_{(1)} = 54,000$; $Y_{(2)} = 89,000$; $Y_{(3)} = 105,000$; $Y_{(4)} = 118,800$; $Y_{(5)} = 126,600$; $Y_{(6)} = 129,000$.

The corresponding graph for the process Y_t is presented in Figure 6 below.

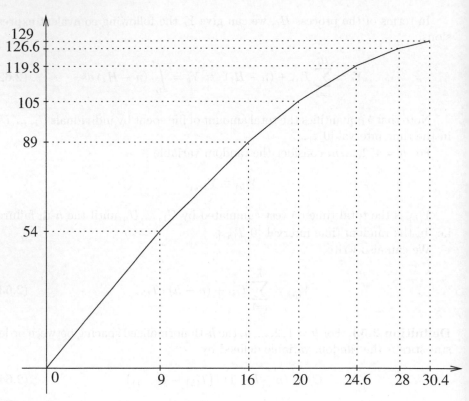

Figure 6.

The following definition is also of interest in the setting of longitudinal observations of failure data.

Given $t_1, ..., t_n$ with $0 \le t_1 \le ... \le t_n$, consider the functions

$$w_1\left(t_1\right) \equiv n \cdot \int_0^{t_1} \lambda_u^{(n)} du$$

$$w_i\left(t_1, ..., t_i\right) \equiv (n - i + 1) \cdot \int_{t_{i-1}}^{t_i} \lambda_u^{(n-i+1)}(t_1, ..., t_{i-1}) du, i = 2, ..., n. \quad (2.66)$$

By using this notation, the expression (2.58) for the joint density $f^{(n)}$ becomes

$$f^{(n)}(t_1, ..., t_n) =$$

$$\lambda_{t_{(1)}}^{(n)} \cdot \prod_{h=2}^{n} \lambda_{t_{(h)}}^{(n-h+1)}(t_{(1)}, ..., t_{(h-1)}) \cdot \exp\left\{ -\sum_{h=1}^{n} w_h\left(t_{(1)}, ..., t_{(h)}\right) \right\}$$

Consider now the random variables defined as follows

$$W_i \equiv w_i\left(T_{(1)}, ..., T_{(i)}\right), i = 1, ..., n. \qquad (2.67)$$

and

$$\Psi_j \equiv \frac{W_1}{n} + ... + \frac{W_j}{n - j + 1}, j = 1, ..., n. \qquad (2.68)$$

Heuristically Ψ_j is the total hazard cumulated by $U_{(j)}$ during its life, where $U_{(j)}$ is the individual failing at the time $T_{(j)}$.

In what follows we derive interesting properties of $(W_1, ..., W_n)$ and $(\Psi_1, ..., \Psi_n)$. To this purpose, for $(u_1, ..., u_n) \in \mathbb{R}_+^n$, we introduce the notation

$$\mathcal{Z}(u_1, ..., u_n) \equiv \left(\frac{u_1}{n}, \frac{u_1}{n} + \frac{u_2}{n - 1}, ..., \sum_{i=1}^{n} \frac{u_i}{n - i + 1}\right) \qquad (2.69)$$

so that we can write

$$(\Psi_1, ..., \Psi_n) = \mathcal{Z}(W_1, ..., W_n) \qquad (2.70)$$

and, from (2.64),

$$\left(T_{(1)}, ..., T_{(n)}\right) = \mathcal{Z}(C_1, ..., C_n) \qquad (2.71)$$

Concerning the sum of coordinates of the transformation $\mathcal{Z}$, note that, $\forall$ $(u_1, ..., u_n) \in \mathbb{R}_+^n$,

$$\frac{u_1}{n} + \left(\frac{u_1}{n} + \frac{u_2}{n - 1}\right) + ... + \left(\sum_{i=1}^{n} \frac{u_i}{n - i + 1}\right) = \sum_{i=1}^{n} u_i. \qquad (2.72)$$

The following fact is well known.

Lemma 2.55. *Let $T_1, ..., T_n$ be exchangeable with a positive density $f^{(n)}(\mathbf{t})$, $\mathbf{t} \in \mathbb{R}_+^n$ and let $\Gamma_n \equiv \left\{\mathbf{t} \in \mathbb{R}_+^n | t_1 \leq t_2 \leq ... \leq t_n\right\}$. The joint density of $\left(T_{(1)}, ..., T_{(n)}\right)$ is given by*

$$\widehat{f}^{(n)}(\mathbf{t}) = n! f^{(n)}(t_1, ..., t_n), \text{ for } \mathbf{t} \in \Gamma_n \qquad (2.73)$$

and

$$\widehat{f}^{(n)}(\mathbf{t}) = 0, \text{ otherwise.}$$

Concerning the joint distribution of the vector $\mathbf{C} \equiv (C_1, ..., C_n)$, we have

Proposition 2.56. **C** *has an absolutely continuous distribution with density*

$$f_{\mathbf{C}}^{(n)}(c_1, ..., c_n) = f^{(n)}\left(\mathcal{Z}(c_1, ..., c_n)\right), (c_1, ..., c_n) \in \mathbb{R}_+^n. \tag{2.74}$$

Proof. The distribution of **C** admits a joint density, since it is defined by means of the one-to-one differentiable transformation (2.64), in terms of $T_{(1)}, ..., T_{(n)}$ which admit a joint density.

The inverse of (2.64) is the transformation $\mathcal{Z}$ in (2.71). Then one has, for $(c_1, ..., c_n) \in \mathbb{R}_+^n$,

$$f_{\mathbf{C}}^{(n)}(c_1, ..., c_n) = \widehat{f}^{(n)}\left(\mathcal{Z}(c_1, ..., c_n)\right) \cdot |J_{\mathcal{Z}}(c_1, ..., c_n)|$$

where $J_{\mathcal{Z}}(c_1, ..., c_n)$ is the Jacobian of the transformation $\mathcal{Z}$. Equation (2.74) is then proved by taking into account (2.73), and by noting that, for $r = 1, ..., n-h$, $|J_{\mathcal{Z}}(c_1, ..., c_n)| = \frac{1}{n!}$, since

$$J_{\mathcal{Z}}(c_1, ..., c_n) = \begin{vmatrix} \frac{1}{n} & 0 & 0 & \cdots & 0 \\ \frac{1}{n} & \frac{1}{n-1} & 0 & \cdots & 0 \\ \frac{1}{n} & \frac{1}{n-1} & \frac{1}{n-2} & \cdots & 0 \\ \cdots & \cdots & \cdots & \cdots & \cdots \\ \frac{1}{n} & \frac{1}{n-1} & \frac{1}{n-2} & \cdots & 1 \end{vmatrix}.$$

$\square$

At this point it is helpful to look at the special case of independent, identically distributed, standard exponential variables. Let then $T_1, ..., T_n$ be such that

$$f^{(n)}(t_1, ..., t_n) = \exp\left\{-\sum_{i=1}^{n} t_i\right\} \tag{2.75}$$

By taking into account Proposition 2.56 and the identity (2.72), we reobtain a well-known result (see e.g. Barlow and Proschan, 1975, p. 60).

Lemma 2.57. $T_1, ..., T_n$ *are independent, identically distributed, standard exponential variables if and only if the corresponding variables* $C_1, ..., C_n$ *are also independent, identically distributed, standard exponential variables.*

By combining this result with the relation (2.71), we also obtain a more general fact: for non-negative random variables $R_1, ..., R_n$, consider the vector $(V_1, ..., V_n)$ defined by

$$(V_1, ..., V_n) = \mathcal{Z}(R_1, ..., R_n)$$

Lemma 2.58. $(V_1, ..., V_n)$ *is distributed as the vector of the order statistics of n independent, identically distributed, standard exponential variables if and only if $R_1, ..., R_n$ are independent, identically distributed, standard exponential variables.*

The above Lemma can be used in the derivation of the following result, concerning the vectors $(W_1, ..., W_n)$ and $(\Psi_1, ..., \Psi_n)$, respectively, defined in (2.67) and (2.68).

Proposition 2.59. *Assume that (2.75) holds for the joint density of $T_1, ..., T_n$. Then $W_1, ..., W_n$ are also independent, identically distributed, standard exponential variables and $(\Psi_1, ..., \Psi_n)$ is distributed as the vector of the order statistics of n independent, identically distributed, standard exponential variables.*

Proof. The condition (2.75) is equivalent to $\lambda_{t_1}^{(n)} = 1, \lambda_t^{(n-h+1)}(t_1, ..., t_{h-1}) = 1$. Whence, in this case, one has

$$w_1(t_1) \equiv n \cdot t_1, \quad w_i(t_1, ..., t_i) \equiv (n - i + 1) \cdot (t_i - t_{i-1}), i = 2, ..., n.$$

or, in other words, by comparing (2.66) with (2.64), we have $(W_1, ..., W_n) = (E_1, ..., E_n)$. From Lemma 2.57 we have that $W_1, ..., W_n$ are i.i.d. standard exponential variables. In order to conclude the proof we only have to take into account (2.70) and to apply Lemma 2.58. $\qquad \square$

A remarkable fact is that the property of $(W_1, ..., W_n)$ and $(\Psi_1, ..., \Psi_n)$, proved in Proposition 2.59 under the condition (2.75), is generally true, as we are going to show next. To this purpose it is helpful to keep in mind the following elementary result.

Let T be a lifetime with a strictly decreasing survival function $\overline{G}$, hazard rate function $r(t)$, and cumulative hazard function $R(t) = \int_0^t r(\xi) d\xi$

Lemma 2.60. *The random variable*

$$V \equiv R(T) = -\log \overline{G}(T)$$

has a standard exponential distribution.

Proof. For $v > 0$,

$$P\{V > v\} = P\{T > R^{-1}(v)\} = \exp\{-R[R^{-1}(v)]\} = \exp\{-v\}.$$

$\qquad \square$

We now turn to considering exchangeable lifetimes $T_1, ..., T_n$, with a positive density $f^{(n)}(\mathbf{t}), \mathbf{t} \in \mathbb{R}_+^n$. Notice that now we are not requiring, for $f^{(n)}$, the condition (2.75); however we can still claim

Proposition 2.61. *a)* $W_1, ..., W_n$ *are i.i.d. with a standard exponential distribution*

b) $(\Psi_1, ..., \Psi_n)$ *is distributed as the vector of the order statistics of n independent, identically distributed, standard exponential variables.*

Proof. As shown by Equation (2.46), $T_{(1)}$ is a non-negative random variable with hazard rate function $n \cdot \lambda_t^{(n)}$. Then W_1 has a standard exponential distribution according to Lemma 2.60.

Now we aim to obtain the conditional distribution of W_2 given W_1. Since $w_1(t_1)$ is a one-to-one mapping, we can equivalently compute the conditional distribution of W_2 given $T_{(1)}$.

By (2.47), the conditional distribution of $T_{(2)} - T_{(1)}$, given $T_{(1)}$, admits the hazard rate function

$$r_{T_{(1)}}(t) \equiv (n-1) \cdot \lambda_{T_{(1)}+t}^{(n-1)}(T_{(1)})$$

Then, by applying again Lemma 2.60

$$\int_0^{T_{(2)}-T_{(1)}} r_{T_{(1)}}(t) dt$$

has a standard exponential distribution, and it is stochastically independent of $T_{(1)}$. On the other hand

$$\int_0^{T_{(2)}-T_{(1)}} r_{t_1}(t) dt = (n-1) \cdot \int_{T_{(1)}}^{T_{(2)}} \lambda_u^{(n-1)}(T_{(1)}) du$$

We can then conclude that $W_2 = w_2\left(T_{(1)}, T_{(2)}\right)$ has a standard exponential distribution, and it is stochastically independent of $W_1 = w_1\left(T_{(1)}\right)$.

Continuing in this way, we obtain that W_h has a standard exponential distribution, and it is stochastically independent of $W_1, ... W_{h-1}$ and a) is proved.

In order to prove b) we simply have to recall Equation (2.70) and then apply Lemma 2.58 to the vector $(W_1, ...W_n)$. $\qquad\square$

A presentation of the result above in a much more general setting has been given in Arjas (1989).

Consider now a strictly positive, exchangeable, and for the rest arbitrary, density $f^{(n)}$ on $\mathbb{R}_+^n$. The proposition 2.61 shows that we can construct a vector of exchangeable lifetimes $T_1, ..., T_n$ with density $f^{(n)}$, starting from a vector of i.i.d. standard exponential variables. Such a construction goes along as follows:

i) fix a vector of i.i.d. standard exponential variables $B_1, ...B_n$.

ii) Consider the transformation $\mathcal{W}$ defined by (2.67) (this is a one-to-one transformation from R_+^n to R_+^n) and denote its inverse by $\mathcal{W}^{-1}$. Note that $\mathcal{W}^{-1}$ depends on the density $f^{(n)}$ by means of the functions w_i's defined by (2.66).

iii) Define a vector of n random variables $\mathbf{T}_{(\cdot)}$ by

$$\mathbf{T}_{(\cdot)} = \mathcal{W}^{-1}(B_1, ...B_n)$$

$\mathbf{T}_{(\cdot)}$ must be distributed like the vector of order statistics of n lifetimes with density $f^{(n)}$.

iv) Define exchangeable variables $T_1, ..., T_n$ by means of a random permutation of coordinates of $\mathbf{T}_{(\cdot)}$. Then $T_1, ..., T_n$ must be distributed according to the density $f^{(n)}$.

Such a construction can be useful in many situations, from the points of view of both theory and practice. It is a special form of the "Total Hazard Construction," defined in general for non-necessarily exchangeable vectors of lifetimes (see Shaked and Shanthikumar, 1994).

2.4.3 M.c.h.r. functions and dynamic sufficiency

Let us start from the analysis of two remarkable special cases: Schur-constant joint densities and m.c.h.r. functions not depending on past failure times, considered in Examples 2.35 and 2.37, respectively. In the first case we have that $\lambda_{t'}^{(n-h)}(t_1, ..., t_h)$ is only a function of the pair (h, y), where h is the number of failures and y is the observed total time on test, $y = \sum_{i=1}^{h} t_i + (n-h)t$; in the second case $\lambda_t^{(n-h)}(t_1, ..., t_h)$ is simply a function of the pair $(h; t)$.

These examples suggest a special definition of sufficiency (*dynamic sufficiency*) for data of the form (2.40). This concept can, in particular, be useful to characterize special probabilistic models for lifetimes; furthermore it can be used to give simpler forms to the transformation $\mathcal{W}^{-1}$, needed for the total hazard construction. Furthermore, conditions of dependence or of multivariate aging (to be studied in the next chapters) are more easily checked or imposed in the case of models characterized by the fact that statistics, of particularly simple form, have the property of dynamic sufficiency.

Denote

$$(0; te) \equiv (0, t, ..., t) \in \{0\} \times R_+^n$$

and, for $0 < t_1 \le ... \le t_h \le t$,

$$(h; t_1, ..., t_h; te) \equiv (h; t_1, ..., t_h; t, ..., t) \in \{h\} \times R_+^n$$

$$\mathcal{G} \equiv \{(h; t_1, ..., t_h; te) | t \ge 0; h = 1, ..., n-1; 0 < t_1 \le ... \le t_h \le t\}$$

$$\widehat{\mathcal{G}} \equiv \mathcal{G} \cup \{(0; te) | t \geq 0\} \tag{2.76}$$

We can look at $\widehat{\mathcal{G}}$ as the space of possible values taken by dynamic histories. Let $q(h; t_1, ..., t_h; te)$ be a measurable function defined on $\widehat{\mathcal{G}}$ (i.e. q is a function of the generic history).

Definition 2.62. q is a dynamic prediction sufficient (d.p.s.) statistic if the following implication holds

$$q(h'; t'_1, ..., t'_{h'}; t'e) = q(h''; t''_1, ..., t''_{h''}; t''e) \Rightarrow$$

$$\lambda_{t'}^{(n-h')}(t'_1, ..., t'_h) = \lambda_{t''}^{(n-h'')}(t''_1, ..., t''_h)$$

The following result generally shows a simple connection between the definitions of prediction sufficiency and dynamic prediction sufficiency.

Proposition 2.63. *Let* $f^{(n)}$ *be such that* $S_m (m = 1, ..., N - 1)$ *is a sequence of prediction sufficient statistics. Then the mapping*

$$q(h; t_1, ..., t_h; t) \equiv (h; S_h(t_1, ..., t_h); t)$$

is a d.p.s.

Proof. By (2.43) and by adapting the formula (1.48) of Chapter 1 to the present notation, we can write

$$\lambda_t^{(n-h)}(t_1, ..., t_h) =$$

$$\frac{\int_t^\infty \cdots \int_t^\infty \phi_{h,n-h}(S_h(t_1, ..., t_h); t, u_{h+2}..., u_n) du_{h+2}...du_n}{\int_t^\infty \cdots \int_t^\infty \phi_{h,n-h}(S_h(t_1, ..., t_h); u_{h+1}, u_{h+2}..., u_n) du_{h+1}...du_n}$$

Then $\lambda_t^{(n-h)}(t_1, ..., t_h)$ only depends on $h, S_h(t_1, ..., t_h)$ and t. $\qquad\square$

Conversely the knowledge of a d.p.s. statistic $w(h; t_1, ..., t_h; te)$ does not give in general any indication about prediction sufficient statistics.

As the formula (2.58) shows, the joint density function depends both on the form of the m.c.h.r. functions and on the values of $t_{(1)}, ..., t_{(n)}$.

Indeed $t_{(1)}, ..., t_{(n)}$ appear in the r.h.s. of (2.58) also as the points between which m.c.h.r. functions are to be integrated. This in general prevents those values from being eliminated when a joint density is divided by a marginal density for computing a conditional distribution. A heuristic explanation of that can also be found in the considerations, developed in the previous subsection, concerning the fact that the system of the m.c.h.r. functions is not an adapt tool to deal with marginal distributions or with conditional distributions, when the conditional event is not a history of "dynamic" type.

2.5   Exercises

Exercise 2.64. Let Θ be a non-negative parameter with a (prior) exponential density

$$\pi_0(\theta) = \beta \exp\{-\beta\theta\}, \ \theta > 0$$

and let T be a lifetime with a conditional hazard rate, given $\Theta = \theta$, $r(t|\theta) = \theta t^\epsilon$ ($\epsilon > -1$), i.e. the conditional distribution of T given Θ is the Weibull distribution, with scale parameter θ and shape parameter $(1 + \epsilon)$. Prove that the predictive hazard rate function is given by

$$r(t) = \int_0^\infty r(t|\theta)\pi(\theta|T > t)d\theta = \frac{t^\epsilon}{\beta + t^{1+\epsilon}}, \ t > 0$$

Hint: use (2.5).

Exercise 2.65. Show that the hazard rate function for the one-dimensional survival function in (2.7) is given by

$$r(t) = r_1(t)P\{\sigma > t|T > t\} + r_2(t)P\{\sigma \le t|T > t\}.$$

Hint: differentiate $\overline{F}(s)$ with respect to s to obtain

$$f(s) = r_2(s) \exp\{-R_2(\sigma)\} \int_0^s \exp\{R_2(\sigma) - R_1(\sigma)\}q(\sigma)d\sigma$$

$$+ r_1(s) \exp\{-R_1(s)\} \int_s^\infty q(\sigma)d\sigma.$$

Exercise 2.66. For a time-transformed exponential model as in Example 2.5, consider the special case when the prior distribution of Θ is Gamma with shape parameter α and scale parameter β. Show that, for D as in (2.18),

$$P\{T_{h+1} > s_{h+1}, ..., T_n > s_n|D\} =$$

$$\left(\frac{\beta + \sum_{j=1}^h R(t_j) + \sum_{j=h+1}^n R(r_j)}{\beta + \sum_{j=1}^h R(t_j) + \sum_{j=h+1}^n R(s_j)} \right)^{(\alpha+h)}$$

Hint: notice that posterior density $\pi(\theta|D)$ is again the density of a Gamma distribution, with parameters

$$\alpha' = \alpha_0 + h, \beta' = \beta_0 + [\sum_{j=1}^h R(t_j) + \sum_{j=h+1}^n R(r_j)]$$

and use the formula (2.31).

Exercise 2.67. For two lifetimes T_1, T_2 in Example 2.19, prove the formula (2.32).

Hint: start proving that

$$\overline{F}^{(2)}(s_1, s_2) =$$

$$\int_0^{s_1} \exp\{-2R_1(\sigma)\} \exp\{-R_2(s_1) - R_2(s_2) + 2R_2(\sigma)]\} q(\sigma) d\sigma +$$

$$\int_{s_1}^{s_2} \exp\{-R_1(s_1) - R_1(\sigma)\} \exp\{-R_2(s_2) + R_2(\sigma)\} q(\sigma) d\sigma +$$

$$\int_{s_2}^{\infty} \exp\{-R_1(s_1) - R_1(s_2)\} q(\sigma) d\sigma.$$

Exercise 2.68. For Example 2.20 write down explicitly the conditional survival probability $P\{T_2 > s_2 | T_1 = t_1, T_2 > r_2\}$.

Exercise 2.69. Prove that, for $1 \leq h \leq n - 1, 0 < t_1 \leq ... \leq t_h \leq t$,

$$\lambda_t^{(n-h)}(t_1, ..., t_h) =$$

$$- \frac{\frac{\partial^{h+1} \overline{F}^{(n)}(s_1, ..., s_h, s_{h+1}, t, ..., t)}{\partial s_1 ... \partial s_{h+1}}\big|_{s_1 = t_1, ..., s_h = t_h, s_{h+1} = t}}{\frac{\partial^h \overline{F}^{(n)}(s_1, ..., s_h, t, ..., t)}{\partial s_1 ... \partial s_h}\big|_{s_1 = t_1, ... s_h = t_h}} \tag{2.77}$$

Exercise 2.70. Prove Lemma 2.42.

Exercise 2.71. (Continuation of Exercise 2.67). Consider again the bivariate survival function in Example 2.19 and let

$$A(s_1, s_2) \equiv r_2(s_1) \exp\{-R_2(s_1) - R_2(s_2)\} \times$$

$$\int_0^{s_1} \exp\{-2R_1(\sigma) + 2R_2(\sigma)\} q(\sigma) d\sigma$$

$$B(s_1, s_2) \equiv r_1(s_1) \exp\{-R_1(s_1) - R_2(s_2)\} \times$$

$$\int_{s_1}^{s_2} \exp\{-R_1(\sigma) + R_2(\sigma)\} q(\sigma) d\sigma$$

$$C(s_1, s_2) \equiv r_1(s_1) \exp\{-R_1(s_1) - R_1(s_2)\} \times$$

$$\int_{s_2}^{+\infty} q(\sigma) d\sigma.$$

Show that

$$\frac{A(s_1, s_2)}{A(s_1, s_2) + B(s_1, s_2) + C(s_1, s_2)} = P\{\sigma \leq s_1 | T_1 = t_1, T_2 > t\}$$

$$\frac{B(s_1, s_2)}{A(s_1, s_2) + B(s_1, s_2) + C(s_1, s_2)} = P\{s_1 < \sigma \leq t | T_1 = t_1, T_2 > t\}$$

$$\frac{C(s_1, s_2)}{A(s_1, s_2) + B(s_1, s_2) + C(s_1, s_2)} = P\{\sigma > t | T_1 = t_1, T_2 > t\}$$

By applying Definition 2.29 then obtain

$$\lambda_t^{(1)}(t_1) = P\{\sigma > t | T_1 = t_1, T_2 > t\}r_1(t) + P\{\sigma \leq t | T_1 = t_1, T_2 > t\}r_2(t)$$

and

$$\lambda_t^{(2)} = P\{\sigma > t | T_1 > t, T_2 > t\}r_1(t) + P\{\sigma \leq t | T_1 > t, T_2 > t\}r_2(t)$$

Exercise 2.72. In the case of two individuals in Example 2.20, where

$$\overline{F}^{(2)}(s_1, s_2) = \sum_{m=0}^{2} p_0(m) P\{T_1 > s_1, T_2 > s_2 | M = m\} =$$

$$p_0(0)\overline{F}_1(s_1)\overline{F}_1(s_2) + \frac{1}{2}p_0(1)\left\{\overline{F}_0(s_1)\overline{F}_1(s_2) + \overline{F}_0(s_2)\overline{F}_1(s_1)\right\} +$$

$$p_0(2)\overline{F}_0(s_1)\overline{F}_0(s_2)$$

show that

$$\lambda_t^{(2)} = -\frac{\frac{\partial \overline{F}^{(2)}(s,t)}{\partial s}|_{s=t}}{\overline{F}^{(2)}(t,t)} =$$

$$\frac{p_0(0)f_1(t)\overline{F}_1(t) + \frac{1}{2}p_0(1)\left\{f_0(t)\overline{F}_1(t) + \overline{F}_0(t)f_1(t)\right\} + p_0(2)f_0(t)\overline{F}_0(t)}{p_0(0)\overline{F}_1(t)\overline{F}_1(t) + \frac{1}{2}p_0(1)\left\{\overline{F}_0(t)\overline{F}_1(t) + \overline{F}_0(t)\overline{F}_1(t)\right\} + p_0(2)\overline{F}_0(t)\overline{F}_0(t)}$$

Hint: Calculate $\frac{\partial \overline{F}^{(2)}(s_1, u)}{\partial s_1}|_{s_1=t_1}$ and take into account Exercise 2.69.

Exercise 2.73. (Continuation). By using Bayes formula, show that the latter expression for $\lambda_t^{(2)}$ is equal to

$$r_1(t)P\{M = 0 | T_1 > t, T_2 > t\} + r_0(t)P\{M = 2 | T_1 > t, T_2 > t\}$$

$$+\frac{1}{2}\{r_1(t) + r_0(t)\}P\{M = 1 | T_1 > t, T_2 > t\}.$$

Exercise 2.74. Show that, in the special case $r(t) \equiv r > 0$, $a(t) \equiv a > 0$, (2.38) becomes (2.36), i.e. the bivariate exponential distribution of Marshall-Olkin, by letting

$$\lambda \equiv a \log(\frac{b + 2r}{b + r}), \lambda' \equiv a \log(\frac{(b + r)^2}{b(b + 2r)}),$$

Exercise 2.75. Check the validity of Equation (2.59).

Exercise 2.76. By applying Proposition 2.48 verify the equation

$$f^{(n)}(t_1, ..., t_n) = \frac{\theta^n}{n!} \exp\{-\theta \cdot t_{(n)}\}$$

for the linear breakdown Ross model in Example 2.50.

Exercise 2.77. Compute the joint density of the bivariate distribution in Example 2.36.

Exercise 2.78. Check the validity of Equation (2.60) at the end of Example 2.51.

2.6 Bibliography

Arjas, E. (1981). The failure and hazard processes in multivariate reliability systems. Math. Oper. Res. 6, 551-562.

Arjas, E. and Norros, I. (1984). Lifelengths and association: a dynamical approach. *Math. Op. Res.* 9, 151-158.

Arjas, E. and Norros, I. (1986). A compensator representation of multivariate life length distributions, with applications. *Scand. J. Stat.* 13, 99-112.

Arjas, E. (1989). Survival models and martingale dynamics. *Scand. J. Statist.*, 16, 177-225.

Aven, T. and Jensen, U. (1999). Stochastic models in reliability. Springer Verlag, New York.

Barlow, R. E. and Irony, T. (1993). The Bayesian approach to quality. In *Reliability and decision making,* R.E. Barlow, C.A. Clarotti and F. Spizzichino, Eds., Chapman & Hall, London.

Barlow, R.E. and Mendel, M.B. (1992). de Finetti-type representations for life distributions. *J. Am. Stat. Soc.,* 87, no. 420, 1116–1122.

Barlow, R. E. and Proschan F. (1975). *Statistical theory of reliability and life testing.* Holt, Rinehart and Winston, New York.

Barlow, R. E. and Proschan F. (1988). Life distributions and incomplete data. In *Handbook of Statistics*, Vol. 7, P.R. Krishnaiah and C.R. Rao Eds., Elsevier, 225-249.

Bergman, B. (1985). On reliability and its applications. *Scand. J. Statist.*, 12, 1-41.

Berliner, L.M. and Hill, B.M. (1988). Bayesian nonparametric survival analysis (with discussion). *J. Am. Stat. Assoc.*, 83, 772-784.

Bremaud, P. (1981). *Point processes and queues. Martingale dynamics.* Springer Verlag, New York.

Caramellino, L. and Spizzichino, F. (1996). WBF property and stochastic monotonicity of the Markov process associated to Schur-constant survival functions. *J. Multiv. Anal.*, 56, 153-163.

Çinlar, E. and Özekici, S. (1987). Reliability of complex devices in random environments. *Prob. Engnr. Inform. Sc.* 1, 97-115.

Çinlar, E. , Shaked, M. and Shantikumar J.G. (1989). On lifetimes influenced by a common environment. *Stoch. Proc. Appl.* 33, 347-359.

Clarotti, C.A. (1992). *L'approche bayesienne predictive en fiabilite': "un noveau regard".* CNES Report N. DCQ 176, Paris.

Cox, D. and Isham, V. (1980). *Point Processes.* Chapman & Hall. London.

Cox, D. and Oakes, D. (1984). *Analysis of survival data.* Chapman & Hall. London.

Diaconis, P. and Freedman, D. (1987). A dozen de Finetti-style results in search of a theory. *Ann. Inst. Henry Poincare* 23, 397-423.

Esary, J.D., Marshall, A.W. and Proschan, F. (1973). Shock models and wear process. *Ann. Probab.*,1, 627-649.

Gaver, D.P. (1963). Random hazard in reliability problems. *Technometrics* 5, 211-226.

Heinrich, G. and Jensen, U. (1996). Bivariate lifetimes distributions and optimal replacement. *Math. Met. Op. Res.*, 44, 31-47.

Hougaard, P. (1987). Modelling multivariate survival. *Scand. J. Statist.* 14, 291-304.

Lawless, J. F. (1982). *Statistical models and methods for lifetime data.* John Wiley & Sons, New York.

Lee, M. T. and Gross, A.J. (1991). Lifetime distributions under unknown environment. *J. Stat. Plann. Inf.* 29, 137-143.

Lindley, D.V. and Singpurwalla, N.D. (1986). Multivariate distributions for the lifelengths of components of a system sharing a common environment. *J. Appl. Prob.*, 23, 418-431.

Martz, H. and Waller, R.A. (1982). *Bayesian reliability analysis.* John Wiley & Sons, New York.

Nappo, G. and Spizzichino, F. (2000). A concept of dynamic sufficiency and optimal stopping of longitudinal observations of lifetimes. *Volume of contributed papers presented at "Mathematical Methods of Reliability", Bordeaux, July 2000.*

Norros, I. (1986). A compensator representation of multivariate life length distributions, with applications. *Scand. J. Statist.* 13, no. 2, 99–112.

Shaked, M. and Shanthikumar, J.G. (1987a). Multivariate hazard rates and stochastic ordering. *Adv. Appl. Probab.* 19, 123-137.

Shaked, M. and Shanthikumar, J.G. (1987b). The multivariate hazard construction. *Stoc. Proc. Appl.*, 24, 85-97.

Singpurwalla, N.D. (1995). Survival in dynamic environments. *Statistical Sciences,* 10, 86-103.

Singpurwalla, N.D. (2000). *Some Cracks in the Empire of Chance.* Private comunication.

Singpurwalla, N.D. and Yougren, M.A. (1993). Multivariate distributions induced by dynamic environments. *Scand. J. Statist.*, 20, 250-261.

Singpurwalla, N.D. and Wilson, S.P. (1995). The exponentiation formula of reliability and survivals: does it always hold?. *Lifetime Data Analysis,* 1, 187-194.

Yashin, A. I. and Arjas, E. (1988). A Note on random intensities and conditional survival functions. *J. Appl. Probab.* 25, 630-635.

Chapter 3

Some concepts of dependence and aging

3.1 Introduction

In this chapter we study some concepts of stochastic orderings, dependence and aging properties for vectors of random quantities.

Such notions provide the essential tools for obtaining inequalities on conditional probability distributions of residual lifetimes, namely for the conditional probabilities of the type

$$P\{T_{h+1} > s_{h+1}, ..., T_n > s_n | T_1 = t_1, ..., T_h = t_h, T_{h+1} > r_{h+1}, ..., T_n > r_n\} \tag{3.1}$$

considered in the previous chapter.

Such inequalities can be useful in various decision problems, as will also be discussed shortly in Chapter 5.

A very rich literature exists concerning the topics of stochastic orderings, dependence and aging properties and related applications in reliability, life-testing and survival analysis; we address the reader to the partial list of references at the end of this chapter for a wider treatment.

The aim of this chapter is just to recall, from these fields, the notions that can be more strictly of use for our purposes.

We shall also discuss some aspects of the concept of dependence in Bayesian analysis. Furthermore we shall consider some preliminaries to the formulation of concepts of aging, for exchangeable lifetimes, in a predictive Bayesian approach.

The fundamental notions are those of univariate and multivariate stochastic orderings. Some of those, for what concerns the univariate case, will be recalled later in this section. Some notions concerning the multivariate case will be briefly illustrated in the next Section 2; a quite exhaustive reference is provided by the volume by Shaked and Shanthikumar (1994). These topics indeed provide the basis for formulating in general definitions and results concerning notions of dependence and of multivariate aging.

In Section 3 we shall consider some concepts of dependence.

We shall then devote Section 4 to discuss some notions of univariate and multivariate aging.

In order to formulate notions of stochastic orderings and dependence it is necessary to consider random vectors $(X_1, ..., X_n)$, in general.

We shall then use the following notation:

for an ordered subset $I \equiv \{i_1, ..., i_{|I|}\} \subset \{1, ..., n\}$ (where $|I|$ is the cardinality of I), $\mathbf{X}_I$ denotes the vector $(X_{i_1}, ..., X_{i_{|I|}})$;

$\mathbf{e}$ denotes a vector with an appropriate number of coordinates all equal to 1, whereas, for $t > 0$, $t\mathbf{e} \equiv (t, ..., t)$; $\tilde{I}$ is the complement of I; finally

$$\mathfrak{h}_t \equiv \{\mathbf{X}_I = \mathbf{x}_I; \mathbf{X}_{\tilde{I}} > t\mathbf{e}\} \tag{3.2}$$

denotes the history

$$\{X_{i_1} = x_{i_1}, ..., X_{i_{|I|}} = x_{i_{|I|}}, X_j > t(j \in \tilde{I})\},$$

where $0 \leq x_{i_1} \leq ... \leq x_{i_{|I|}} \leq t$.

For a vector of lifetimes $\mathbf{X}$, for a history $\mathfrak{h}_t$ as in (3.2) and for $j \in \tilde{I}$, the symbol

$$\lambda_{j|I}(t|x_{i_1}, ..., x_{i_{|I|}})$$

is used to denote the multivariate conditional hazard rate (see Shaked and Shanthikumar, 1994):

$$\lambda_{j|I}(t|x_{i_1}, ..., x_{i_{|I|}}) = \lim_{\Delta t \to 0} \frac{1}{\Delta t} P\{X_j \leq t + \Delta t | \mathfrak{h}_t\}.$$

In general, given a history $\mathfrak{h}_t$, we need to consider the joint distribution of the vector of residual lifetimes $(X_j - t)$, $j \in \tilde{I}$, conditional on the observation of $\mathfrak{h}_t$, which will be denoted by

$$\mathcal{L}[(\mathbf{X}_{\tilde{I}} - t\mathbf{e})|\mathfrak{h}_t].$$

Sometimes one must refer to observed histories $\mathfrak{h}_t$ without specifying which is the set I of indexes of individuals failed along $\mathfrak{h}_t$. In such cases the notation $(\mathbf{X} - t\mathbf{e})^+$ is used to denote the vector whose components are the residual lifetimes for surviving individuals and 0 for failed individuals, so that it is possible to write, for all $t > 0$,

$$\mathbf{X} = (\mathbf{X} \wedge t\mathbf{e}) + (\mathbf{X} - t\mathbf{e})^+,$$

where

$$(\mathbf{X} \wedge t\mathbf{e}) \equiv (X_1 \wedge t, ..., X_n \wedge t).$$

When we condition with respect to a history $\mathfrak{h}_t$ as above (where events of null probability can possibly appear in the conditioning) we shall assume the existence of probability density functions for the joint distribution of $\mathbf{X}$. It can also happen that this is tacitly assumed some other times, when clear from the context. When considering equations or inequalities for conditional probabilities of the form (3.1), it will be also understood that they are valid with probability one (i.e. up to possible events of probability 0).

Our interest is in the condition of exchangeability and we use special notation when dealing with such a specific case: as done in most of the last chapter, the symbol $\mathbf{T} \equiv (T_1, ..., T_n)$ denotes a vector of exchangeable lifetimes and the symbols

$$\lambda_t^{(n)}, \lambda_t^{(n-h)}(t_1, ..., t_h)$$

are respectively used in place of $\lambda_{j|\emptyset}(t)$ and $\lambda_{j|I}(t|t_1, ..., t_h)$, for arbitrary j, I, such that $j \notin I$ and $|I| = h(1 \leq h < n)$.

Throughout this chapter several examples will be presented to illustrate how various concepts manifest in different exchangeable models for lifetimes.

As a background for the arguments to be presented henceforth, in the next subsection we recall basic definitions about notions of orderings between pairs of one-dimensional probability distributions.

3.1.1 One-dimensional stochastic orderings

We start with a very short presentation about *totally positive functions of order 2*, in the simplest form possible (see Karlin, 1968, for more general definitions, results, and proofs).

Let $K(x, y)$ be a non-negative function of the two variables x, y ($x \in \mathcal{X} \subset \mathbb{R}$, $y \in \mathcal{Y} \subset \mathbb{R}$)

Definition 3.1. $K(x, y)$ is *totally positive of order 2* (in short TP_2) if, for $y < y'$, the ratio

$$\frac{K(x, y)}{K(x, y')}$$

is a decreasing function of x.

Proposition 3.2. *(Basic composition formula). Let $K(x, y)$ and $H(y, z)$ be two TP_2 functions ($x \in \mathcal{X}, y \in \mathcal{Y}, z \in \mathcal{Z} \subset \mathbb{R}$) and let μ be a positive measure over $\mathcal{Y}$. Then the function*

$$L(x, z) \equiv \int_{\mathcal{Y}} K(x, y) H(y, z) d\mu(y)$$

is TP_2.

Let $K(x,y)$ $(x \in \mathcal{X}, y \in \mathcal{Y})$ be a TP_2 function and let $h(y)$ be a real function. For a given measure μ, consider the function $g(x)$ defined by the integral

$$g(x) \equiv \int_{\mathcal{Y}} K(x,y)h(y)d\mu(y)$$

Proposition 3.3. *(Variation sign diminishing property). If $h(y)$ changes sign at most once, then $g(x)$ changes sign at most once.*

Now we recall, and comment on, three well-known types of stochastic ordering for one-dimensional distributions, namely *likelihood ratio ordering*, *hazard rate ordering* and *usual stochastic ordering*.

We prefer in the beginning to formulate definitions for the case of probability distributions concentrated on $R_+ \equiv [0,+\infty)$ and admitting a density.

Let then $\overline{F}_1$ and $\overline{F}_2$ be two one-dimensional survival functions ($\overline{F}_1(0) = \overline{F}_2(0) = 1$) and denote by f_i and r_i the corresponding density functions and failure rate functions, respectively ($i = 1, 2$):

$$\overline{F}_i(t) = exp\{-\int_0^t r_i(\xi)d\xi\}, f_i(t) = r_i(t)exp\{-\int_0^t r_i(\xi)d\xi\}, \text{for } t \geq 0 \quad (3.3)$$

Definition 3.4. $\overline{F}_1$ is less than $\overline{F}_2$ in the usual stochastic ordering, written $\overline{F}_1 \leq_{st} \overline{F}_2$, if

$$\overline{F}_1(t) \leq \overline{F}_2(t), \forall t \geq 0.$$

Definition 3.5. $\overline{F}_1$ is less than $\overline{F}_2$ in the hazard rate ordering, written $\overline{F}_1 \leq_{hr} \overline{F}_2$, if

$$r_1(t) \geq r_2(t), \forall t \geq 0.$$

Definition 3.6. $\overline{F}_1$ is less than $\overline{F}_2$ in the likelihood ratio ordering, written $\overline{F}_1 \leq_{lr} \overline{F}_2$, if $f_i(t)$ is TP_2 in i and t, namely

$$\frac{f_1(t')}{f_2(t')} \geq \frac{f_1(t'')}{f_2(t'')}, \text{ for } 0 \leq t' \leq t''.$$

Example 3.7. Let us consider two exponential distributions with parameter λ_1 and λ_2, respectively, i.e.

$$\overline{F}_i(t) = \exp\{-\lambda_i t\}, r_i(t) = \lambda_i,$$

It is trivial that $\lambda_1 > \lambda_2$ implies $\overline{F}_1 \leq_{lr} \overline{F}_2$, $\overline{F}_1 \leq_{rh} \overline{F}_2$ and $\overline{F}_1 \leq_{st} \overline{F}_2$.

Example 3.8. For $t > 0$, let

$$r_1(t) = \max(t, 1), \ r_2(t) = 1.$$

Then

$$f_1(t) = \exp\{-t\}, \text{ for } 0 \leq t \leq 1; \ f_1(t) = e^{-1/2} t \exp\{-\frac{t^2}{2}\}, \ \text{for } t > 1$$

whence

$$\frac{f_1(t)}{f_2(t)} = 1, \text{ for } 0 \leq t \leq 1; \ \frac{f_1(t)}{f_2(t)} = e^{-1/2} t \exp\{-\frac{t^2}{2} + t\}, \text{ for } t > 1$$

It is $\overline{F}_1 \leq_{hr} \overline{F}_2$ and $\overline{F}_1 \leq_{st} \overline{F}_2$, but the ratio $\frac{f_1(t)}{f_2(t)}$ is not everywhere decreasing.

Example 3.9. Let

$$r_1(t) = 2, \text{ for } 0 \leq t \leq 4 \text{ and } t \geq 5; \ r_1(t) = 0, \text{ for } 4 < t < 5$$

and

$$r_2(t) = 1, \forall t \geq 0.$$

As is immediately seen, $\overline{F}_1 \leq_{st} \overline{F}_2$ but neither $\overline{F}_1 \leq_{lr} \overline{F}$ nor $\overline{F}_1 \leq_{hr} \overline{F}_2$ hold true.

The following result is obvious.

Theorem 3.10. $\overline{F}_1 \leq_{hr} \overline{F}_2$ *is equivalent to the condition*

$$\overline{F}_i(t) \ TP_2 \ \text{in the variables } i \text{ and } t.$$

The following two results are of fundamental importance and are well known.

Theorem 3.11. $\overline{F}_1 \leq_{lr} \overline{F}_2 \Rightarrow \overline{F}_1 \leq_{hr} \overline{F}_2 \Rightarrow \overline{F}_1 \leq_{st} \overline{F}_2.$

Proof. For $v > u \geq 0, \overline{F}_1 \leq_{lr} \overline{F}_2$ implies

$$\int_u^v \int_v^\infty f_1(t) \, f_2(t') \, dt dt' \geq \int_v^\infty \int_u^v f_1(t) \, f_2(t') \, dt dt'$$

that is

$$\overline{F}_2(v) \int_u^v f_1(t) \, dt \geq \overline{F}_1(v) \int_u^v f_2(t) \, dt$$

or, equivalently,

$$\overline{F}_2(v)\left[\overline{F}_1(u) - \overline{F}_1(v)\right] \geq \overline{F}_1(v)\left[\overline{F}_2(u) - \overline{F}_2(v)\right]$$

which is just the condition $\overline{F}_i$ TP_2 in the variables i and t.

On the other hand $r_1(t) \geq r_2(t)$ implies $\int_0^t r_1(\xi)d\xi \geq \int_0^t r_2(\xi)d\xi$, and then

$$\overline{F}_2(t) = \exp\{-\int_0^t r_1(\xi)d\xi\} \leq \exp\{-\int_0^t r_2(\xi)d\xi\} = \overline{F}_2(t)$$

$$\square$$

Let T_1 and T_2 be two random variables with survival functions $\overline{F}_1$ and $\overline{F}_2$, respectively and let $\psi:\mathbb{R}_+ \to \mathbb{R}$ be a given function. Then the expected value of $\psi(T_i)$ $(i = 1, 2)$, when it does exist, is denoted by

$$\mathbb{E}[\psi(T_i)] = \int_0^\infty \psi(t)dF_i(t).$$

Theorem 3.12. $\overline{F}_1 \leq_{st} \overline{F}_2$ *if and only if*

$$\mathbb{E}[\psi(T_1)] \leq \mathbb{E}[\psi(T_2)]$$

for any non-decreasing function ψ such that both the expected values exist.

Proof. For simplicity's sake we first consider the case when $F_i(t) = 1 - \overline{F}_i(t)$ $(i = 1, 2)$ is everywhere continuous and strictly increasing over $[0, \infty)$, so that $F_i^{-1} : [0, 1) \to [0, \infty)$ is also continuous and strictly increasing.

Let U be a random variable uniformly distributed over the interval $[0, 1]$ and consider the random variables

$$X = F_1^{-1}(U), Y = F_2^{-1}(U).$$

It is easy to check that the survival functions of X and Y coincide with $\overline{F}_1$ and $\overline{F}_2$, respectively and then we can write

$$\mathbb{E}[\psi(T_1)] = \mathbb{E}[\psi(X)], \mathbb{E}[\psi(T_2)] = \mathbb{E}[\psi(Y)].$$

Furthermore, by the assumption $\overline{F}_1 \leq_{st} \overline{F}_2$, we have $F_1^{-1} \leq F_2^{-1}$, whence $X \leq Y$ and then

$$\mathbb{E}[\psi(X)] \leq \mathbb{E}[\psi(Y)].$$

When $F_i(t)$ $(i = 1, 2)$ is not strictly increasing we can take as F_i^{-1} the right continuous inverse of F_i, defined by

$$F_1^{-1}(u) = \sup\{x|F_i(x) \leq u\}, \forall u \in (0, 1).$$

$$\square$$

Theorems 3.12 and 3.10 show that the definitions of $\overline{F}_1 \leq_{st} \overline{F}_2$ and $\overline{F}_1 \leq_{hr} \overline{F}_2$ can immediately be extended, in a natural manner, to the arbitrary case of a pair of distributions which are not necessarily both absolutely continuous.

Even the definition of likelihood ratio ordering can be extended to the general case, by considering the densities of the two distributions with respect to some measure which dominates both of them (such a measure always exists) and by requiring monotonicity of the ratio between two such densities.

It is also clear that there is no reason to limit the above definitions to pairs of one-dimensional distributions concentrated on $[0, +\infty)$.

When convenient, random quantities of concern will also be denoted by symbols $X, Y, \ldots$

Even though a stochastic ordering $\leq_*$ (where $*$ stands for st, hr, or lr) is an ordering relation between two probability distributions, one can also write $X \leq_* Y$. This means that the distribution of X is majorized by the distribution of Y in the $*$ sense.

3.1.2 Stochastic monotonicity and orderings for conditional distributions

We are often interested in comparing different one-dimensional distributions which arise as conditional distributions for a same scalar random quantity X, given different observed events.

In this respect we can obtain concepts of *stochastic monotonicity* corresponding in a natural way to the notions of stochastic orderings.

Let X be given and let Z be a random variable taking values in the domain $\mathcal{Z} \subset \mathbb{R}^d$ for some $d = 1, 2, \ldots$.

We assume that the joint distribution of (X, Z) is absolutely continuous and consider the conditional distributions of X given events of the type $\{Z = z\}$, with $z \in \mathcal{Z}$; let $f_X(\cdot|z)$ and $\overline{F}_X(\cdot|z)$ in particular denote the conditional density and survival function, respectively.

Let $\overset{\sim}{\prec}$ be a given partial ordering defined on $\mathcal{Z}$ and let $\leq_*$ be a fixed one-dimensional ordering (thus $* = st, hr$, or lr).

Definition 3.13. X is *stochastically increasing in Z in the $*$ ordering with respect to $\overset{\sim}{\prec}$*, if the following implication holds:

$$z' \overset{\sim}{\prec} z'' \Rightarrow \overline{F}_X(\cdot|z') \leq_* \overline{F}_X(\cdot|z'').$$

When no partial ordering $\overset{\sim}{\prec}$ on $\mathbb{R}^d$ is specified, one tacitly refers to the natural partial ordering on $\mathbb{R}^d$ (of course we have a total ordering in the case $d = 1$).

We now translate some fundamental properties of the univariate $\leq_{lr}$ ordering into properties for conditional distributions in a number of different cases. This permits us to underline some aspects of interest in Bayesian statistics.

Simple examples of applications will be presented in Chapter 5.

Remark 3.14. (Preservation of the $\leq_{lr}$ ordering under *posterization*). Let Θ be a scalar parameter with a prior density $\pi(\theta)$ and a (non-necessarily scalar) statistical observation X with a likelihood $f(x|\theta)$.

The posterior density of Θ given the observation $\{X = x\}$ is then given by

$$\pi(\theta|x) \propto \pi(\theta)f(x|\theta).$$

Compare now the two posterior densities $\pi_1(\theta|x)$ and $\pi_2(\theta|x)$, corresponding to two different prior densities $\pi_1(\theta)$ and $\pi_2(\theta)$:

$$\pi_i(\theta|x) \propto \pi_i(\theta)f(x|\theta), i = 1, 2.$$

It is obvious that, for x such that $f(x|\theta) > 0$ for all θ, the following implication holds:

$$\pi_1(\theta) \leq_{lr} \pi_2(\theta) \Rightarrow \pi_1(\theta|x) \leq_{lr} \pi_2(\theta|x)$$

Remark 3.15. (Creation of $\leq_{lr}$ ordering by means of *posterization*). Consider a scalar parameter Θ with a given prior density $\pi(\theta)$ and a scalar statistical observation X with a likelihood $f(x|\theta)$.

Assume $f(x|\theta)$ to be a TP_2 function, i.e. let X be increasing in Θ in the sense of the $\leq_{lr}$ ordering, so that the conditional distribution of X, given $\{\Theta = \theta'\}$ is less than the conditional distribution of X, given $\{\Theta = \theta''\}$ in the $\leq_{lr}$ ordering, if $\theta' < \theta''$.

It is readily seen that this is equivalent to the TP_2 property of the posterior density $\pi(\theta|x)$ of Θ given $\{X = x\}$ i.e. to the condition that Θ is increasing in X in the $\leq_{lr}$ ordering: if we compare the effects of two different results $\{X = x'\}$ and $\{X = x''\}$, with $x' < x''$, then

$$\pi(\theta|x') \leq_{lr} \pi(\theta|x'').$$

Compare now the posterior distributions for Θ given two different results of the form $\{X > x'\}$ and $\{X > x''\}$, again with $x' < x''$: for

$$\overline{F}(t|\theta) = \int_t^{+\infty} f(\xi|\theta)d\xi,$$

we have

$$\pi(\theta|X > x') \propto \pi(\theta) \cdot \overline{F}(x'|\theta), \pi(\theta|X > x'') \propto \pi(\theta) \cdot \overline{F}(x''|\theta).$$

It is then obvious that the comparison

$$\pi(\theta|X > x') \leq_{lr} \pi(\theta|X > x'')$$

can be obtained by requiring the weaker condition that the function $\overline{F}(x|\theta)$ is TP_2, namely that X is increasing in Θ in the $\leq_{hr}$ ordering. Note that the same condition also implies

$$\pi(\theta|X = x) \leq_{lr} \pi(\theta|X > x), \text{ for any } x.$$

Remark 3.16. ($\leq_{lr}$ ordering *for predictive* distributions). Consider a scalar parameter Θ and a scalar statistical observation X with a TP_2 likelihood $f(x|\theta)$.

Compare the two predictive densities $g_1(x)$ and $g_2(x)$, corresponding to two different prior densities $\pi_1(\theta)$ and $\pi_2(\theta)$:

$$g_i(x) = \int_{-\infty}^{\infty} \pi_i(\theta) \cdot f(x|\theta)d\theta, i = 1, 2.$$

As an immediate consequence of the basic composition formula, we get the following implication:

$$\pi_1(\theta) \leq_{lr} \pi_2(\theta) \Rightarrow g_1(x) \leq_{lr} g_2(x).$$

3.2 Multivariate stochastic orderings

In this section we shall discuss definitions of multivariate concepts of stochastic orderings which extend the notions seen above for the one-dimensional case; this leads to four different concepts:
a) Usual multivariate stochastic ordering
b) Multivariate likelihood ratio ordering
c) Cumulative hazard rate ordering
d) Multivariate hazard rate ordering

3.2.1 Usual multivariate stochastic ordering

This concept is very well known. For two d-dimensional random vectors $\mathbf{X}, \mathbf{Y}$ we want to define the condition $\mathbf{X} \leq_{st} \mathbf{Y}$.

A subset $U \subset \mathbb{R}^n$ is an upper set (or an increasing set) if its indicator function is increasing (with respect to the natural partial ordering $\leq$), i.e.

$$\mathbf{x}' \in U, \mathbf{x}' \leq \mathbf{x}'' \Rightarrow \mathbf{x}'' \in U.$$

Definition 3.17. $\mathbf{X}$ is stochastically smaller than $\mathbf{Y}$ (written $\mathbf{X} \leq_{st} \mathbf{Y}$) if, for any upper set U, one has

$$P\{\mathbf{X} \in U\} \leq P\{\mathbf{Y} \in U\}.$$

Remark 3.18. The condition $\mathbf{X} \leq_{st} \mathbf{Y}$ trivially implies the inequality

$$F_{\mathbf{X}}(\mathbf{x}) \geq F_{\mathbf{Y}}(\mathbf{x}),$$

for any $\mathbf{x} \in R^n$. If $n > 1$, the vice versa is not true, as well-known examples show.

A direct multidimensional analog of Theorem 3.12, shows however that Definition 3.17 can be seen as the most natural multidimensional extension of Definition 3.4:

Theorem 3.19. $\mathbf{X} \leq_{st} \mathbf{Y}$ *if and only if*

$$\mathbb{E}[\psi(\mathbf{X})] \leq \mathbb{E}[\psi(\mathbf{Y})]$$

for any non-decreasing function $\psi{:}\mathbb{R}^n \to \mathbb{R}$ *such that both the expected values exist.*

Proof. It is obvious that the condition $\mathbb{E}[\psi(\mathbf{X})] \leq \mathbb{E}[\psi(\mathbf{Y})]$ for any non-decreasing function ψ is sufficient for $\mathbf{X} \leq_{st} \mathbf{Y}$, since

$$P\{\mathbf{X} \in U\} = \mathbb{E}[\mathbf{1}_U(\mathbf{X})], P\{\mathbf{Y} \in U\} = \mathbb{E}[\mathbf{1}_U(\mathbf{Y})]$$

and, U being an increasing set, the indicator function $\mathbf{1}_U$ is increasing.

On the other hand, any increasing function $\psi{:}R^n \to R$ can be approximated as follows:

$$\psi(\mathbf{x}) = \lim_{m \to \infty} \psi_m(\mathbf{x})$$

where ψ_m has the form

$$\psi_m(\mathbf{x}) = \sum_{i=1}^{m} = a_{m,i} \mathbf{1}_{U_{m,i}}(\mathbf{x}) - b_m,$$

and $a_{m,i} > 0$ $(i = 1, ..., m)$, $U_{m,i}$ are upper sets of $\mathbb{R}^n$ and $b_m \in \mathbb{R}^n$. Then it is

$$\mathbb{E}[\psi(\mathbf{X})] = \lim_{m \to \infty} \mathbb{E}[\psi_m(\mathbf{X})] = \lim_{m \to \infty} \sum_{i=1}^{m} a_{m,i} P\{\mathbf{X} \in U_{m,i}\} - b_m$$

$$\mathbb{E}[\psi(\mathbf{Y})] = \lim_{m \to \infty} \mathbb{E}[\psi_m(\mathbf{Y})] = \lim_{m \to \infty} \sum_{i=1}^{m} a_{m,i} P\{\mathbf{Y} \in U_{m,i}\} - b_m$$

Then, if $\mathbf{X} \leq_{st} \mathbf{Y}$,

$$\mathbb{E}[\psi(\mathbf{X})] \leq \mathbb{E}[\psi(\mathbf{Y})].$$

$\square$

3.2.2 Multivariate likelihood ratio ordering

Also the notion of lr ordering admits the following natural extension to the multidimensional case (see Karlin and Rinott, 1980; Whitt, 1982).

For two n-dimensional random vectors $\mathbf{X}, \mathbf{Y}$ with joint densities $f_{\mathbf{X}}$ and $f_{\mathbf{Y}}$, the condition $\mathbf{X} \leq_{lr} \mathbf{Y}$ is defined as follows:

Definition 3.20. $\mathbf{X}$ is smaller than $\mathbf{Y}$ in the likelihood ratio (written $\mathbf{X} \leq_{lr} \mathbf{Y}$) if, for any pair of vectors $\mathbf{u}, \mathbf{v} \in \mathbb{R}^n$

$$f_{\mathbf{Y}}(\mathbf{u} \vee \mathbf{v}) f_{\mathbf{X}}(\mathbf{u} \wedge \mathbf{v}) \geq f_{\mathbf{X}}(\mathbf{u}) f_{\mathbf{Y}}(\mathbf{v}) \tag{3.4}$$

where

$$\mathbf{u} \wedge \mathbf{v} \equiv (u_1 \wedge v_1, ..., u_n \wedge v_n), \ \mathbf{u} \vee \mathbf{v} \equiv (u_1 \vee v_1, ..., u_n \vee v_n).$$

Remark 3.21. Note that, by taking $\mathbf{u} \geq \mathbf{v}$, from the definition of $\leq_{lr}$, we immediately obtain

$$\frac{f_{\mathbf{X}}(\mathbf{v})}{f_{\mathbf{Y}}(\mathbf{v})} \geq \frac{f_{\mathbf{X}}(\mathbf{u})}{f_{\mathbf{Y}}(\mathbf{u})}. \tag{3.5}$$

In the one-dimensional case such a condition just amounts to state $\mathbf{X} \leq_{lr} \mathbf{Y}$. In the multivariate case, on the contrary, the two condition are in general not equivalent.

As it is easy to show, the condition (3.5), however, implies that $\mathbf{X} \leq_{lr} \mathbf{Y}$ if $\mathbf{X}$ is MTP_2, according to the next Definition 3.38 (see Kochar, 1999, p. 351); the condition implies that $\mathbf{X} \leq_{st} \mathbf{Y}$ if $\mathbf{X}$ is *associated*, according to the next Definition 3.37 (see Shaked and Shanthikumar, 1994, p. 118).

A fundamental property is the following result, which shows that the $\leq_{lr}$ order is maintained under conditioning upon a suitable class of positive probability events (see e.g. Whitt, 1982 or Shaked and Shanthikumar, 1994, Theorem 4.E.1).

The set $A \subset \mathbb{R}^n$ is said to be a *lattice* if the following implication holds:

$$\theta, \theta' \in A \Rightarrow \theta \wedge \theta' \in A, \theta \vee \theta' \in A. \tag{3.6}$$

Theorem 3.22. *Let* $\mathbf{X} \leq_{lr} \mathbf{Y}$ *and let* A *be a subset with the lattice property and such that* $P\{\mathbf{X} \in A\} > 0, P\{\mathbf{Y} \in A\} > 0$. *Then*

$$\mathbf{X} | \mathbf{X} \in A \leq_{lr} \mathbf{Y} | \mathbf{Y} \in A.$$

A further fundamental property of the $\leq_{lr}$ order lies in that it is maintained when passing to marginal distributions: if $\mathbf{X} \leq_{lr} \mathbf{Y}$ and $I \subset \{1, ..., n\}$ then $\mathbf{X}_I \leq_{lr} \mathbf{Y}_I$.

3.2.3 Multivariate hazard rate and cumulative hazard rate orderings

Here we recall the definitions of two other concepts of multivariate stochastic ordering, namely the multivariate hazard rate ordering and the cumulative hazard rate ordering.

In order to define them we have to look at vectors of non-negative random quantities, thought of as lifetimes of individuals $U_1, ..., U_n$.

Their probability distributions are assumed to be absolutely continuous.

Let $\mathbf{Y}$ and $\mathbf{X}$ be two n-dimensional vectors of life-times as above and let $\lambda_{.|.}(\cdot|\cdot)$ and $\eta_{.|.}(\cdot|\cdot)$ be the corresponding multivariate hazard functions.

Definition 3.23. $\mathbf{X}$ is smaller than $\mathbf{Y}$ in the *multivariate hazard rate ordering* (written $\mathbf{X} \leq_{hr} \mathbf{Y}$) if $\forall u > 0$,

$$\eta_{r|J}(u|\mathbf{x}_J) \geq \lambda_{r|I}(u|\mathbf{y}_I),$$

whenever $I \subset J$, $r \notin J$, $x_i \leq y_i$ for all $i \in I$.

In order to define the cumulative hazard ordering the following two definitions are needed.

Definition 3.24. For a dynamic history $\mathfrak{h}_t \equiv \{\mathbf{X}_I = \mathbf{x}_I, \mathbf{X}_{\tilde{I}} > te\}$, where $I \equiv \{i_1, ..., i_{|I|}\}$, $x_{i_1} \leq x_{i_2} ... \leq x_{i_{|I|}}$, the *cumulative hazard* of a component $i \in \tilde{I}$ at time t is defined by

$$\Psi_{i|I}(t|\mathbf{x}_I) = \int_0^{x_{i_1}} \lambda_{i|\emptyset}(u)du+$$

$$\sum_{j=2}^{|I|} \int_{x_{i_{j-1}}}^{x_{i_j}} \lambda_{i|\{i_1,...,i_{j-1}\}}(u|x_{i_1}, x_{i_2}, ..., x_{i_{j-1}})du + \int_{x_{i_{|I|}}}^t \lambda_{i|I}(u|\mathbf{x}_I)du$$

Definition 3.25. For two dynamic histories

$$\mathfrak{h}_t \equiv \{\mathbf{X}_I = \mathbf{y}_I, \mathbf{X}_{\tilde{I}} > te\}, \quad \mathfrak{h}'_s \equiv \{\mathbf{X}_J = \mathbf{x}_J, \mathbf{X}_{\tilde{J}} > se\}$$

with $\mathbf{x}_I < te$, $\mathbf{y}_J < se$, we say that $\mathfrak{h}_t$ *is less severe than* $\mathfrak{h}'_s$ whenever

$$0 < t \leq s, \ I \subset J, \ \mathbf{x}_I \leq \mathbf{y}_I.$$

Let $\mathbf{X} \equiv (X_1, ..., X_n)$ and $\mathbf{Y} \equiv (Y_1, ..., Y_n)$ be two n-dimensional vectors of lifetimes and let $\Phi_{.|.}(\cdot|\cdot)$ and $\Psi_{.|.}(\cdot|\cdot)$ be the corresponding cumulative hazard functions.

Definition 3.26. $\mathbf{X}$ is smaller than $\mathbf{Y}$ in the *cumulative hazard rate ordering* (written $\mathbf{X} \leq_{ch} \mathbf{Y}$) if $\forall t > 0$,

$$\Phi_{r|J}(t|\mathbf{x}_J) \geq \Psi_{r|I}(t|\mathbf{y}_I)$$

whenever $I \subset J, r \notin J, x_i \leq y_i$ for all $i \in I$, i.e. when

$$\mathfrak{h}_t \equiv \{\mathbf{Y}_I = \mathbf{y}_I, \mathbf{Y}_{\tilde{I}} > t\mathbf{e}\}$$

is less severe than

$$\mathfrak{h}'_t \equiv \{\mathbf{X}_J = \mathbf{x}_J, \mathbf{Y}_{\tilde{J}} > t\mathbf{e}\}.$$

3.2.4   Some properties of multivariate stochastic orderings and examples

Remark 3.27. Consider the special case when both the vectors $\mathbf{X} \equiv (X_1, ..., X_n)$ and $\mathbf{Y} \equiv (Y_1, ..., Y_n)$ have stochastically independent components.

Denote by $\overline{F}_i$ the one-dimensional survival functions of Y_i $(i = 1, ..., n)$ and by $\overline{G}_i$ the one-dimensional survival functions of X_i $(i = 1, ..., n)$, respectively.

Assume that $\overline{F}_i$ and $\overline{G}_i$, respectively, admit density functions f_i, g_i. Denote by r_i, q_i their hazard rate functions and by R_i, Q_i their cumulative hazard functions.

Which are the meanings of the above notions of multivariate ordering in such a case?

One can easily check the following:

$\mathbf{X} \leq_{lr} \mathbf{Y}$ if and only if $X_i \leq_{lr} Y_i, i = 1, ..., n$.

$\mathbf{X} \leq_{hr} \mathbf{Y}$ if and only if $q_i(u) \geq r_i(u), \forall u \geq 0, i = 1, ..., n$.

$\mathbf{X} \leq_{ch} \mathbf{Y}$ if and only if $R_i(u) \geq Q_i(u), \forall u \geq 0, i = 1, ..., n$.

$\mathbf{X} \leq_{st} \mathbf{Y}$ if and only if $X_i \leq_{st} Y_i, i = 1, ..., n$.

Note, however, that the condition $R_i(u) \geq Q_i(u), \forall u \geq 0$ is equivalent to $X_i \leq_{st} Y_i$.

Then the two conditions $\mathbf{X} \leq_{ch} \mathbf{Y}$ and $\mathbf{X} \leq_{st} \mathbf{Y}$ are equivalent in this special case.

A fundamental fact, in general, is the following chain of implications.

Theorem 3.28. *(See Shaked and Shanthikumar, 1994). Let $\mathbf{Y}$ and $\mathbf{X}$ be two non-negative random vectors with absolutely continuous distributions.*

Then the condition $\mathbf{X} \leq_{lr} \mathbf{Y}$ implies $\mathbf{X} \leq_{hr} \mathbf{Y}$. $\mathbf{X} \leq_{hr} \mathbf{Y}$ implies $\mathbf{X} \leq_{ch} \mathbf{Y}$. $\mathbf{X} \leq_{ch} \mathbf{Y}$ implies $\mathbf{X} \leq_{st} \mathbf{Y}$.

Remark 3.29. The above chain of implications is analogous to the one holding for univariate orderings where, however, the two concepts $X \leq_{ch} Y$ and $X \leq_{st} Y$ do coincide.

When $\mathbf{X}$ and $\mathbf{Y}$ are not non-negative vectors, one can in any case establish that $\mathbf{X} \leq_{lr} \mathbf{Y}$ implies $\mathbf{X} \leq_{st} \mathbf{Y}$.

The definitions given so far will now be illustrated by means of examples and counter-examples, taken from exchangeable models considered in Chapter 2.

Example 3.30. ($\leq_{hr}$ and $\leq_{lr}$ comparisons). In Example 2.37 of Chapter 2, we saw that, for the special Ross model characterized by a system of m.c.h.r.'s of the form

$$\lambda_t^{(n-h)}(t_1, ..., t_h) = \frac{\theta}{n-h},$$

the corresponding joint density is given by $f^{(n)}(t_1, ..., t_n) = \frac{\theta^n}{n!} \exp\{-\theta \cdot t_{(n)}\}$.

Consider now two such models corresponding to a pair of different values θ and $\widehat{\theta}$, with $0 < \theta < \widehat{\theta}$, and compare $f^{(n)}(\mathbf{t})$ with $\widehat{f}^{(n)}(\mathbf{t})$ and $\{\lambda_t^{(k)}(\cdot)\}$ with $\{\widehat{\lambda}_t^{(k)}(\cdot)\}$.

Let $\mathbf{T}$ and $\widehat{\mathbf{T}}$ denote two vectors of random lifetimes with joint density $f^{(n)}(\mathbf{t})$ and $\widehat{f}^{(n)}(\mathbf{t})$, respectively. It is easy to check that $\widehat{\mathbf{T}} \leq_{hr} \mathbf{T}$ and $\widehat{\mathbf{T}} \leq_{lr} \mathbf{T}$.

Example 3.31. (Comparisons in cases of negative dependence). Let $\mathbf{T} \equiv (T_1, T_2)$ have an exchangeable joint density characterized by the set of m.c.h.r. functions

$$\lambda_t^{(2)} = \frac{\theta + a}{2}, \lambda_t^{(1)}(t_1) = \theta,$$

where $a > 0$. When $a > \theta$, we have a case of negative dependence.

Consider now $\widehat{\mathbf{T}} \equiv (T_1, T_2)$ with a joint distribution defined similarly but with

$$\widehat{\lambda}_t^{(2)} = \frac{\widehat{\theta} + a}{2}, \lambda_t^{(1)}(t_1) = \widehat{\theta},$$

where $\widehat{\theta} > \theta$. Since $\lambda_t^{(2)} < \widehat{\lambda}_t^{(2)}$ and $\lambda_t^{(1)}(t_1) < \widehat{\lambda}_t^{(1)}(t_1)$, $\forall 0 \leq t \leq t_1$, we can wonder whether it is $\widehat{\mathbf{T}} \leq_{hr} \mathbf{T}$.

This is not necessarily the case, since the condition $\widehat{\mathbf{T}} \leq_{hr} \mathbf{T}$ also requires that $\lambda_t^{(2)} \leq \widehat{\lambda}_t(t_1)$, $\forall \, 0 < t_1 \leq t$, and this condition is not satisfied if $a > 2\widehat{\theta} - \theta$.

Denote by $\widehat{F}_{R,\pi_0}$ the joint predictive survival function of $\mathbf{T} \equiv (T_1, ..., T_n)$ in the statistical model of "proportional hazards" considered in Examples 2.18 and 2.40.

Θ is a non-negative parameter with a prior density $\pi_0(\theta)$ and $T_1, ..., T_n$ are conditionally i.i.d given Θ, with a conditional distribution of the form $G(s|\theta) = \exp\{-\theta \cdot R(s)\}$; i.e.

$$\widehat{F}_{R,\pi_0}(t_1, ..., t_n) \equiv \int_0^\infty \exp\{-\theta \cdot \sum_{i=1}^n R(t_i)\}\pi_0(\theta)d\theta. \tag{3.7}$$

We saw that, given a history $\mathfrak{h}_t \equiv \{\mathbf{T}_I = \mathbf{t}_I, \mathbf{T}_{\bar{I}} > t\mathbf{e}\}$, the posterior density of Θ is given by

$$\pi(\theta|\mathfrak{h}_t) \propto \theta^h \exp\{-\theta[\sum_{i=1}^{h} R(t_i) + (n-h)R(t)]\}\pi_0(\theta)$$

and the multivariate conditional hazard rate for a surviving individual U_j is

$$\lambda_j(t|\mathfrak{h}_t) = \lambda_t^{(n-h)}(t_1, ..., t_h) = r(t) \cdot E(\Theta|\mathfrak{h}_t), \qquad (3.8)$$

where $r(t) \equiv R'(t)$.

In what follows we show that, by comparing distributions of the form (3.7), we can get further examples and counter-examples concerning notions of multivariate stochastic orderings.

Example 3.32. ($\leq_{hr}$ comparison). Consider two different initial densities $\pi_0(\theta)$ and $\tilde{\pi}_0(\theta)$ with $\tilde{\pi}(\theta) \leq_{lr} \pi_0(\theta)$. By using arguments similar to those in Remark 3.14, we see that, for any history $\mathfrak{h}_t$,

$$\tilde{\pi}(\theta|\mathfrak{h}_t) \leq_{lr} \pi(\theta|\mathfrak{h}_t),$$

where

$$\tilde{\pi}(\theta|\mathfrak{h}_t) \propto \theta^h \exp\{-\theta[\sum_{i=1}^{h} R(t_i) + (n-h)R(t)]\} \cdot \tilde{\pi}_0(\theta).$$

On the other hand, by Remark 3.15, we can also see that if we consider a different history $\mathfrak{h}'_t$, with $\mathfrak{h}_t$ less severe than $\mathfrak{h}'_t$, then we obtain

$$\pi(\theta|\mathfrak{h}_t) \leq_{lr} \pi(\theta|\mathfrak{h}'_t), \tilde{\pi}(\theta|\mathfrak{h}_t) \leq_{lr} \tilde{\pi}(\theta|\mathfrak{h}'_t)$$

whence

$$\tilde{\pi}(\theta|\mathfrak{h}_t) \leq_{lr} \pi(\theta|\mathfrak{h}'_t).$$

By taking into account the expression (3.8) for the multivariate conditional hazard rate, we can conclude

$$\widehat{F}_{R,\pi_0} \leq_{hr} \widehat{F}_{R,\tilde{\pi}_0}$$

Example 3.33. Here we continue the example above, considering the special case of a comparison between gamma distributions for Θ; more precisely we take

$$\pi_0(\theta) \propto 1_{[\theta>0]}\theta^{\alpha-1} \exp\{-\beta \cdot \theta\}$$

and

$$\widetilde{\pi}_0(\theta) \propto 1_{[\theta>0]}\theta^{\alpha-1} \exp\{-\beta' \cdot \theta\}.$$

In order to get the condition $\widetilde{\pi}(\theta) \leq_{lr} \pi_0(\theta)$, we set $\beta' > \beta$.

By specializing the formula (2.29) of Chapter 2, we see that, with the above choice for $\pi_0(\theta)$ and $\widetilde{\pi}_0(\theta)$, the joint density functions of $\widehat{F}_{R,\pi_0}$ and $\widehat{F}_{R,\widetilde{\pi}_0}$ respectively are

$$f_{R,\pi_0}(t_1,...,t_n) \propto \frac{1}{[\beta + \sum_{i=1}^{n} R(t_i)]^{\alpha+n}},$$

$$f_{R,\widetilde{\pi}_0}(t_1,...,t_n) \propto \frac{1}{[\beta' + \sum_{i=1}^{n} R(t_i)]^{\alpha+n}}$$

It can be easily checked that the inequality (3.4) holds and then $\widehat{F}_{R,\pi_0} \leq_{lr} \widehat{F}_{R,\widetilde{\pi}_0}$.

Example 3.34. For an arbitrarily fixed prior probability density π_0 over $[0, +\infty)$, and two absolutely continuous cumulative hazard functions $R(s)$, $\widetilde{R}(s)$ consider the two joint distributions identified by $\widehat{F}_{R,\pi_0}$ and $\widehat{F}_{\widetilde{R},\pi_0}$. One can check (see Exercise 3.84) that the following implication holds true:

$$R(s) \geq \widetilde{R}(s) \Rightarrow \widehat{F}_{R,\pi_0} \leq_{st} \widehat{F}_{\widetilde{R},\pi_0}.$$

However the inequality $R(s) \geq \widetilde{R}(s)$ does not necessarily guarantee the condition $\widehat{F}_{R,\pi_0} \leq_{hr} \widehat{F}_{\widetilde{R},\pi_0}$. Indeed, consider again $\pi_0(\theta) \propto 1_{[\theta>0]}\theta^{\alpha-1} \exp\{-\beta \cdot \theta\}$ and the histories $\mathfrak{h}_t = \mathfrak{h}'_t \equiv \{\mathbf{T}_I = \mathbf{t}_I, \mathbf{T}_{\bar{I}} > te\}$. We can write, with obvious meaning of notation,

$$\lambda_j(t|\mathfrak{h}_t) = \lambda_t^{(n)} = \rho(t) \cdot \mathbb{E}(\Theta|\mathfrak{h}_t) = \frac{\rho(t)}{\beta + n \cdot R(t)}$$

$$\widetilde{\lambda}_j(t|\mathfrak{h}_t) = \widetilde{\lambda}_t^{(n)} = \widetilde{\rho}(t) \cdot \mathbb{E}(\Theta|\mathfrak{h}_t) = \frac{\widetilde{\rho}(t)}{\beta + n \cdot \widetilde{R}(t)}.$$

Of course we can find functions R and $\widetilde{R}$ such that $R \geq \widetilde{R}$ and

$$\frac{\rho(t)}{\beta + n \cdot R(t)} < \frac{\widetilde{\rho}(t)}{\beta + n \cdot \widetilde{R}(t)},$$

for some $t > 0$.

3.3   Some notions of dependence

Informally, a notion of dependence for a random vector $\mathbf{X} \equiv (X_1, ..., X_n)$ can be looked at as follows: it is a property concerning stochastic orderings and stochastic monotonicity for conditional distributions of a set of coordinates, given the observation of various types of events concerning $\mathbf{X}$.

Dependence conditions, on the other hand, may give rise to inequalities between different distributions having the same set of (one-dimensional) marginal distributions. The concept of *copula* has fundamental importance in this respect (see e.g. Nelsen, 1999).

Actually a property of dependence is one of the probability distributions of $\mathbf{X}$.

Denote by $\Delta_\mathcal{D}$ the class of all probability distributions on $\mathbb{R}^n$ which share a fixed dependence property $\mathcal{D}$.

Since we usually identify the distribution of $\mathbf{X}$ by means of its joint survival function F, it will be also written $F \in \Delta_\mathcal{D}$ to mean that the random vector $\mathbf{X}$ satisfies the property $\mathcal{D}$.

3.3.1   Positive dependence

In what follows we recall the definitions of some concepts of positive dependence.

Let us start with those which are mostly well known.

Definition 3.35. $\mathbf{X}$ is *positively correlated*, written $F \in \Delta_{PC}$, if

$$Cov(X_i, X_j) \geq 0$$

for any pair $1 \leq i \neq j \leq n$, provided the covariances do exist.

Definition 3.36. X is *positively upper orthant dependent*, written $F \in \Delta_{PUOD}$, if

$$P\{\bigcap_{i=1}^{n}(X_i > x_i)\} \geq \prod_{i=1}^{n} P\{(X_i > x_i)\}$$

for any choice of $x_i \in \mathbb{R}$ $(1 \leq i \leq n)$.

In other words $F \in \Delta_{PUOD}$ if and only if random vectors of the form $(\mathbf{1}_{(X_1 > x_1)}, ..., \mathbf{1}_{(X_n > x_n)})$ are positively correlated. Similarly, one defines the property $F \in \Delta_{PLOD}$ (*positively lower orthant dependent*).

Definition 3.37. $\mathbf{X}$ is *associated*, written $F \in \Delta_{Assoc}$, if, for any pair of nondecreasing functions ϕ, ψ ($\phi, \psi : \mathbb{R}^n \to \mathbb{R}$), one has that the pair $(\phi(\mathbf{X}), \psi(\mathbf{X}))$ is positively correlated.

Some other notions of positive dependence can directly originate from notions of stochastic ordering.

Consider in particular the multivariate stochastic orderings $\leq_{lr}$, $\leq_{hr}$, and $\leq_{ch}$ defined in the previous section.

None of these is actually an ordering in the common sense; indeed none of them has the property of being *reflexive*:

we can easily find random vectors $\mathbf{X}$ which do not satisfy $\mathbf{X} \leq_{lr} \mathbf{X}, \mathbf{X} \leq_{hr}$ $\mathbf{X}$, or $\mathbf{X} \leq_{ch} \mathbf{X}$.

The condition

$$\mathbf{X} \leq_* \mathbf{X},$$

where $*$ stands for the symbols lr, or hr, or ch, is a condition of positive dependence for $\mathbf{X}$ and this then leads us to the three different concepts defined next.

Definition 3.38. (Karlin and Rinott, 1980). The density of $\mathbf{X}$ is *multivariate totally positive* of order 2 written $F \in \Delta_{MTP_2}$, if $\mathbf{X} \leq_{lr} \mathbf{X}$, namely if F is absolutely continuous and its density is such that

$$f(\mathbf{u} \wedge \mathbf{v}) f(\mathbf{u} \vee \mathbf{v}) \geq f(\mathbf{u}) f(\mathbf{v}) \qquad (3.9)$$

Definition 3.39. (Shaked and Shanthikumar, 1987). A non-negative random vector $\mathbf{X}$ is *hazard rate increasing upon failure*, written $F \in \Delta_{HIF}$, if $\mathbf{X} \leq_{hr} \mathbf{X}$ namely if

$$\lambda_{i|I}(t|\mathbf{x}_I) \leq \lambda_{i|J}(t|\mathbf{y}_J)$$

whenever $I \subset J, i \notin J, t \cdot \mathbf{e}_I > \mathbf{x}_I \geq \mathbf{y}_I, \mathbf{y}_{J-I} \leq t \cdot \mathbf{e}_{J-I}$.

And finally, in an analogous way, the following definition can be given (see Norros, 1986; Shaked and Shanthikumar, 1990).

Definition 3.40. A non-negative random vector $\mathbf{X}$ has *supportive lifetimes*, written $F \in \Delta_{SL}$, if $\mathbf{X} \leq_{ch} \mathbf{X}$.

In order to define a last concept of positive dependence, we consider a non-negative random vector $\mathbf{X}$ and compare two different histories $\mathfrak{h}_t$ and $\mathfrak{h}'_t$ of the form

$$\mathfrak{h}_t \equiv \{\mathbf{X}_I = \mathbf{x}_I; \mathbf{X}_{\tilde{I}} > t\mathbf{e}\}, \qquad (3.10)$$

$$\mathfrak{h}'_t \equiv \{\mathbf{X}_I = \mathbf{x}_I, \mathbf{X}_i = t; X_{\tilde{I}-\{i\}} > t\mathbf{e}\} \qquad (3.11)$$

where $i \in \tilde{I}$

Definition 3.41. A non-negative random vector $\mathbf{X}$ is *weakened by failures*, written $F \in \Delta_{WBF}$, if

$$\mathcal{L}[(\mathbf{X}_{\tilde{I}-\{i\}} - t\mathbf{e})|\mathbf{X}_I = \mathbf{t}_I, X_i = t, \mathbf{X}_{\tilde{I}-\{i\}} > t\mathbf{e}] \leq_{st}$$

$$\leq_{st} \mathcal{L}[(\mathbf{X}_{\tilde{I}-\{i\}} - t\mathbf{e})|X_I = \mathbf{t}_I, \mathbf{X}_{\tilde{I}} > t\mathbf{e}]$$

For more details, see Shaked and Shanthikumar (1990); see also Arjas and Norros (1984) and Norros (1985) for different formulations of this concept.

Among the notions listed so far there is a chain of implications explained by the following result.

Theorem 3.42. *For a non-negative random vector $\mathbf{X}$ with an absolutely continuous distribution, one has*

$$MTP_2 \Rightarrow HIF \Rightarrow SL \Rightarrow WBF \Rightarrow$$

$$Association \Rightarrow PUOD \Rightarrow Positive\ Correlation.$$

Proof. The chain of implications $MTP_2 \Rightarrow HIF \Rightarrow SL$ is a direct consequence of Definitions 3.37, 3.38, 3.39 and Theorem 3.28.

The implication $SL \Rightarrow WBF$ can be seen as a corollary of Theorem 3.43 below.

The implication $WBF \Rightarrow Association$ is valid under the condition of absolute continuity (see Arjas and Norros, 1984).

The implication $Association \Rightarrow PUOD$ is an obvious and very well-known consequence of the definition of Association.

The implication $PUOD \Rightarrow Positive\ Correlation$ is achieved by taking into account the identities

$$E(X_i \cdot X_j) = \int_0^\infty \int_0^\infty P\{X_i > u, X_j > v\} du dv$$

for $1 \leq i \neq j \leq n$ and

$$E(X_i) = \int_0^\infty P\{X_i > u\} du.$$

$\square$

The following result shows that conditions of MTP$_2$, HIF, and SL can be characterized in a way which is formally similar to the definition of WBF.

Theorem 3.43. *(Shaked and Shanthikumar, 1990). Let* $\mathbf{X}$ *be a non-negative random vector with absolutely continuous distribution. Then*

  a) $\mathbf{X}$ *is SL if and only if*

$$\mathcal{L}[(\mathbf{X}_{\tilde{I}-\{i\}} - te)|\mathbf{X}_I = \mathbf{t}_I, \mathbf{X}_i = t, \mathbf{X}_{\tilde{I}-\{i\}} > te] \leq_{ch} \mathcal{L}[(\mathbf{X}_{\tilde{I}-\{i\}} - te)|\mathbf{X}_I = \mathbf{t}_I, \mathbf{X}_{\tilde{I}} > te]$$

  b) $\mathbf{X}$ *is HIF if and only if*

$$\mathcal{L}[(\mathbf{X}_{\tilde{I}-\{i\}} - te)|\mathbf{X}_I = \mathbf{t}_I, \mathbf{X}_i = t, \mathbf{X}_{\tilde{I}-\{i\}} > te] \leq_{hr} \mathcal{L}[(\mathbf{X}_{\tilde{I}-\{i\}} - te)|\mathbf{X}_I = \mathbf{t}_I, \mathbf{X}_{\tilde{I}} > te]$$

  c) $\mathbf{X}$ *is* MTP_2 *if and only if*

$$\mathcal{L}[(\mathbf{X}_{\tilde{I}-\{i\}} - te)|\mathbf{X}_I = \mathbf{t}_I, \mathbf{X}_i = t, \mathbf{X}_{\tilde{I}-\{i\}} > te] \leq_{lr} \mathcal{L}[(\mathbf{X}_{\tilde{I}-\{i\}} - te)|\mathbf{X}_I = \mathbf{t}_I, \mathbf{X}_{\tilde{I}} > te]$$

Also for the concepts of positive dependence presented in this section, we can get concrete examples by analyzing, once again, the special exchangeable models introduced in Chapter 2.

Example 3.44. (Proportional hazard models). Consider a proportional hazard model $\overline{F}_{R,\pi_0}$. This is MTP_2 for any choice of the initial density π_0 as an application of the Theorem 3.59 below.

Example 3.45. (Schur-constant densities). Let $\mathbf{T} \equiv (T_1, ..., T_n)$ have a Schur-constant density: $f^{(n)}(\mathbf{t}) = \phi_n(\sum_{i=1}^n t_i)$ for a suitable function $\phi_n : \mathbb{R}_+ \to \mathbb{R}_+$.

When $\phi_n(t) = \int_0^{+\infty} \theta^n \exp\{-\theta t\}\pi_0(\theta)d\theta$, this is a proportional hazard model $\overline{F}_{R,\pi_0}$, with $R(t) = t$, and then it is MTP_2.

On the other hand note that $\mathbf{T}$ is MTP_2 if and only if

$$\phi_n\left(\sum_{i=1}^n (x_i \vee y_i)\right) \phi_n\left(\sum_{i=1}^n (x_i \wedge y_i)\right) \geq \phi_n\left(\sum_{i=1}^n x_i\right) \phi_n\left(\sum_{i=1}^n y_i\right)$$

Then, since $\sum_{i=1}^n x_i \vee y_i - \sum_{i=1}^n x_i = \sum_{i=1}^n y_i - \sum_{i=1}^n x_i \wedge y_i$, we see that $\mathbf{X}$ is MTP_2 if and only if ϕ_n is log-convex.

As a matter of fact, the latter condition is a necessary condition for $\mathbf{T}$ being WBF (see Caramellino and Spizzichino, 1996). Since $MTP_2 \Rightarrow HIF \Rightarrow SL \Rightarrow WBF$, we see that, in this special case, the properties MTP_2, HIF, SL, and WBF are equivalent.

Example 3.46. (Hazards depending on the number of survivals). Consider the case of an exchangeable vector $\mathbf{T}$ where, for any $I \subset \{1, 2, ..., n\}$, $i \notin I$, $\lambda_{i|I}(t)$ is a quantity non-depending on i and depending on I only through its cardinality $|I| = n - h$:

$$\lambda_t^{(n-h)}(x_1, ..., x_h) = \lambda_{i|I}(t) = \varphi_h(t).$$

We can of course expect that $\mathbf{T}$ satisfy some condition of positive dependence when $\varphi_h(t)$ is a non-decreasing function of h. By applying b) in Theorem. 3.43, one can actually check that, in such a case, $\mathbf{T}$ is HIF.

Example 3.47. Here we go back to the case of units undergoing a change of stress level at a random time σ. By Proposition 2.43, we can get

$$\lambda_t^{(n-h)}(t_1, ..., t_h) =$$

$$r_1(t) \cdot P\{\sigma > t | T_{(1)} = t_1, ..., T_{(h)} = t_h, T_{(h+1)} > t\} +$$

$$+r_2(t) \cdot P\{\sigma \leq t | T_{(1)} = t_1, ..., T_{(h)} = t_h, T_{(h+1)} > t\}.$$

Under the condition $r_1(t) \leq r_2(t)$ we have

$$\lambda_t^{(n-h)}(t_1, ..., t_h) \leq \lambda_t^{(n-h')}(t_1', ..., t_{h'}')$$

for two different observed histories

$$\mathfrak{h}_t \equiv \{T_{(1)} = t_1, ..., T_{(h)} = t_h, T_{(h+1)} > t\}$$

$$\mathfrak{h}_t' \equiv \{T_{(1)} = t_1', ..., T_{(h')} = t_{h'}', T_{(h'+1)} > t\},$$

if and only if

$$P\{\sigma > t | \mathfrak{h}_t\} \geq P\{\sigma > t | \mathfrak{h}_t'\} \tag{3.12}$$

We now want to show that $\mathbf{X}$ is WBF. We must then check that the inequality (3.12) holds when

$$\mathfrak{h}_t \equiv \{T_{(1)} = t_1, ..., T_{(h)} = t_h, T_{(h+1)} > t\}$$

$$\mathfrak{h}_t' \equiv \{T_{(1)} = t_1, ..., T_{(h)} = t_h, T_{(h+1)} = t_{h+1}, T_{(h+2)} > t\}.$$

This can be seen by using arguments similar to those of Remark 3.15 and applying Theorem 3.63 below (Exercise 3.86).

It is worth remarking that checking the WBF property directly requires lengthy computations even in the simplest case when σ is an exponential time and the two subsequent stress levels do not depend on the age of units, namely $r_1(t) = \lambda_1, r_2(t) = \lambda_2$.

Remark 3.48. If σ were known, then the m.c.h.r functions $\lambda_t^{(n-h)}(t_1, ..., t_h)$ in the example above would be clearly equal to $r_1(t)$ or to $r_2(t)$ according to whether $\sigma > t$ or $\sigma \leq t$. Then, conditionally on σ, $\lambda_t^{(n-h)}(t_1, ..., t_h | \sigma)$ would neither depend on $t_1, ..., t_h$ nor on h; indeed the lifetimes $T_1, ..., T_n$ are conditionally independent (identically distributed) given σ.

Example 3.49. (Common mode failures). Consider the case of lifetimes $T_1, ..., T_n$ where $T_j = V_j \wedge U$ with $V_1, ..., V_n$ i.i.d. lifetimes and U a random time independent of $V_1, ..., V_n$, denoting the time to the arrival of a shock. Of course we expect that $T_1, ..., T_n$ are positively dependent, due to the common dependence on U. It is easy to see that they are associated, as a consequence of Properties 2 and 4 in Remark 3.50 below.

Remark 3.50. The concept of association has important structural properties, whose applications in the reliability analysis of *coherent systems* are very well known (see Barlow and Proschan, 1975). The same properties hold, however, also for the stronger concept of MTP$_2$.

In particular, letting the symbol * stand for "associated" or for "MTP$_2$" (* is seen as a condition on a random vector or on a set of random variables), one has

Property 1. Any single random variable is *.

Property 2. The union of independent * random vectors is *.

Property 3. A subset of a * set of random variables is * (i.e., the property * is maintained for marginal distributions).

Property 4. If $\mathbf{X} \equiv (X_1, ..., X_n)$ is * and $\psi_1, ..., \psi_m : \mathbb{R}^n \to \mathbb{R}$ are increasing functions, then $(\psi_1(\mathbf{X}), ..., \psi_m(\mathbf{X}))$ is *.

Remark 3.51. As self-evident, there are some basic differences between the notions of positive correlation, PUOD, association and MTP$_2$ on one side, and notions of WBF, SL, and HIF on the other side. The latter notions are of "dynamic type", i.e. they are formulated in terms of the conditional distribution of residual lifetimes $(X_j - t), j \in \tilde{I}$, given an observed history of the form (3.2) with $(I \subset \{1, 2, ..., n\})$. These notions, which are essentially considered only for the case of non-negative random vectors, have been defined in the frame of reliability theory and survival analysis and their interest has been essentially limited to such a context.

On the contrary the other notions are of interest in all fields of application of probability. It is interesting to remark that, in particular, the notions of MTP$_2$ and association have been dealt with (under different terminology and almost independently up to a certain extent) in the two different fields of reliability and mechanical statistics. The implication MTP$_2 \Rightarrow$ Association dates back to Sarkar (1969), Fortuin, Ginibre and Kasteleyn (1971), and Preston (1974), see also the book by Ligget (1985).

3.3.2 Negative dependence

Here we mention a few definitions of negative dependence. Also for this case, many definitions have been presented in the literature; we shall only consider some which may be seen as the negative counterparts of Definitions 3.36, 3.37, 3.38 and, finally, the "condition N".

Examples of exchangeable distributions which exhibit negative dependence properties will be given in some of the exercises at the end of this chapter.

Definition 3.52. $X_1, ..., X_n$ are *negatively upper orthant dependent*, written $F \in \Delta_{NUOD}$, if

$$P\{\bigcap_{i=1}^{n}(X_i > x_i)\} \leq \prod_{i=1}^{n} P\{(X_i > x_i)\}$$

for any choice of $x_i \in R(1 \leq i \leq n)$.

In other words $F \in \Delta_{NUOD}$ if and only if random vectors of the form $(1_{(X_1>x_1)}, ..., 1_{(X_n>x_n)})$ are negatively correlated.

$X_1, ..., X_n$ are *negatively lower orthant dependent*, written $F \in \Delta_{NLOD}$, if

$$P\{\bigcap_{i=1}^{n}(X_i < x_i)\} \leq \prod_{i=1}^{n} P\{(X_i < x_i)\}$$

$X_1, ..., X_n$ are *negatively orthant dependent*, written $F \in \Delta_{NOD}$, if they are both *NUOD* and *NLOD*.

Definition 3.53. (Joag-Dev and Proschan, 1983). $X_1, ..., X_n$ are negatively associated, written $F \in \Delta_{NA}$, if for disjoint subsets I and J of $\{1, ..., n\}$ and increasing functions ϕ and ψ,

$$Cov(\phi(\mathbf{X}_I), \psi(\mathbf{X}_J)) \leq 0.$$

Definition 3.54. (Karlin and Rinott, 1980b). Let $X_1, ..., X_n$ admit a joint density function f. f is *multivariate reverse regular of order 2* (MRR_2), if

$$f(\mathbf{u} \wedge \mathbf{v})f(\mathbf{u} \vee \mathbf{v}) \leq f(\mathbf{u})f(\mathbf{v}) \tag{3.13}$$

A different concept which is similar to, but weaker than, the latter definition is reverse regular of order 2 in pairs.

Definition 3.55. (Block, Savits and Shaked, 1982). $\mathbf{X} \equiv (X_1, ..., X_n)$ is RR_2 *in pairs* if, for any $1 \leq i \neq j \leq n$,

$$P\{a \leq X_i \leq b, c \leq X_j \leq d\} \cdot P\{a' \leq X_i \leq b', c' \leq X_j \leq d'\} \leq$$

$$\leq P\{a \leq X_i \leq b, c' \leq X_j \leq d'\} \cdot P\{a \leq X_i \leq b, c' \leq X_j \leq d'\}$$

whenever $b < a', d < c'$.

A stronger notion is, on the contrary, given in the following

Definition 3.56. (Block, Savits and Shaked, 1982). $\mathbf{X} \equiv (X_1, ..., X_n)$ satisfies *condition* N, written $F \in \Delta_N$, if there exist a real number s and $n + 1$ independent random variables $S_1, ..., S_{n+1}$, each with a log-concave density (or PF_2 probability function) such that

$$(X_1, ..., X_n) =_{st} [(S_1, ..., S_{n+1})| \sum_{i=1}^{n} S_i = s].$$

See the papers by Ebrahimi and Ghosh (1981); Karlin and Rinott (1980); Block, Savits and Shaked (1982) and Joag-Dev and Proschan (1983), for several properties of the above concepts and for further definitions of negative dependence.

In any case, some fundamental properties are the following:

Property 1. A set of independent random variables is *, where * stands for $NA, MRR_2, NUOD, NLOD$.

Property 2. A subset of * random variables is *, where * stands for NA, $NUOD, NLOD$.

Property 3. The union of independent sets of * variables is *, where * stands for $NA, MRR_2, NUOD, NLOD$.

Property 4. If $X_1, ..., X_n$ are * and $\psi_1, ..., \psi_n$ are increasing functions then

$$\psi_1(T_1), ..., \psi_n(T_n)$$

are *, where * stands for NA, RR_2 in pairs, $NUOD, NLOD$. Actually for NA something stronger happens: increasing functions defined on disjoint subsets of a set of NA variables are NA.

3.3.3 Simpson-type paradoxes and aspects of dependence in Bayesian analysis

The present subsection will be devoted to the discussion of some aspects, concerning dependence, which arise when probability is understood in a subjective sense. These aspects will also be relevant in our discussion about concepts of multivariate aging.

First we notice that exchangeable models are likely to manifest relevant properties of dependence: often cases of finite extendibility arise from negative, symmetric dependence (think for instance of the case in Example 2.4 of Chapter 2) while cases of infinite extendibility, i.e. cases of conditionally i.i.d quantities, may manifest some form of positive dependence (in this respect see Theorems 3.57-3.59 below).

In general, we point out that the kind of dependence among variables of interest is sensible to the state of information of the observer about the same variables.

As already observed, dependence can be created as an effect of uncertainty about some relevant quantities. In particular variables which are conditionally independent given a quantity Θ, are actually stochastically dependent in their (unconditional) distribution, unless the distribution of Θ is degenerate. In Examples 3.44 and 3.47 above we found properties of positive dependence for conditionally i.i.d. lifetimes with conditional densities $g(t|\theta)$ satisfying certain conditions of stochastic monotonicity in θ.

These provide instances of the following general fact: *when $X_1, ..., X_n$ are conditionally independent given a finite-dimensional parameter Θ, we can get positive dependence properties, by combining stochastic monotonicity conditions on the conditional distributions $G_i(x|\theta)$ with positive dependence conditions on Θ (no such condition is needed when Θ is a scalar).*

Some results in this direction are the following.

Theorem 3.57. *(Jogdeo, 1978). Let Θ be an associated vector and let $G_i(x|\theta)$ be stochastically increasing in θ, with respect to the usual stochastic order. Then $(X_1, ..., X_n)$ is associated.*

Theorem 3.58. *(Shaked and Spizzichino, 1998). Let Θ be a scalar and let $G_i(x|\theta)$ be absolutely continuous and stochastically increasing in θ, with respect to the one-dimensional hazard rate order. Then $(X_1, ..., X_n)$ is WBF.*

Theorem 3.59. *(Shaked and Spizzichino, 1998). Let Θ be a scalar and let $G_i(x|\theta)$ be absolutely continuous and stochastically increasing in θ, with respect to the one-dimensional likelihood ratio order. Let also the support of $(X_1, ..., X_n)$ be a lattice. Then $(X_1, ..., X_n)$ is MTP_2.*

Theorems 3.58 and 3.59 can be extended to the case when Θ is multidimensional, by adding suitable MTP_2 properties for the joint distribution (see Shaked and Spizzichino, 1998).

On this theme see also Çinlar, Shaked and Shanthikumar (1989) and Szekli (1995).

Exercises 3.87, 3.88, 3.94 provide examples related to the above results.

Remark 3.60. In conclusion we see that cases of positive dependence arise from situations of conditional independence, given some relevant parameters, and are due just to lack of knowledge about the same parameters.

We are particularly interested in the cases of lifetimes $T_1, ..., T_n$, which, conditionally on Θ, are identically distributed, besides being independent; this is then the case when $G_i(\cdot|\theta)$ is independent of the index i. Notice that in such situations the one-dimensional marginal of the predictive survival function

$$\overline{F}^{(n)}(t_1, ..., t_n) = \int \overline{G}(t_1|\theta)...\overline{G}(t_n|\theta)d\Pi_0(\theta)$$

is the mixture-type survival function

$$\overline{F}^{(1)}(t) = \int \overline{G}(t|\theta)d\Pi_0(\theta).$$

This will have an effect on the study of multivariate aging properties of $\overline{F}^{(n)}$ (see Subsection 4.3). Notice also that these situations of positive dependence can be viewed as conceptually different from other cases of positive dependence such as the one considered in Example 3.46 above (see also Remark 2.41 in Section 3 of Chapter 2).

Consider now cases which are more general than conditional independence; again we may want to compare dependence properties of conditional distributions (given the values of relevant parameters) with dependence properties of the corresponding predictive (i.e. unconditional) distributions.

In some cases situations of lack of knowledge about parameters may create some positive dependence; in other cases situations of lack of knowledge may destroy some types of dependence present in the conditional distributions.

When the latter happens we say that we meet a "*Simpson-type* paradox"; i.e. we say that a Simpson-type paradox appears when the conditional law of a random vector $\mathbf{X}$ exhibits a dependence property for every possible value of the conditioning random vector Θ, but does not exhibit the same property unconditionally. More formally, let $\mathcal{D}$ be a property of dependence, $\Delta_{\mathcal{D}}$ the family of joint distributions with the property $\mathcal{D}$, and let $(\mathbf{X}, \Theta)$ be a pair of random vectors; the paradox can then be described as follows:

$$\mathcal{L}(\mathbf{X}|\Theta = \theta) \in \Delta_D, \quad \forall \theta \in L, \text{ and } \mathcal{L}(\mathbf{X}) \notin \Delta_D, \qquad (3.14)$$

where L is the support of Θ.

In the language of Samuels (1993), this is an *association distortion phenomenon*. For more details and references about Simpson-type paradoxes see Scarsini and Spizzichino (1999); see also Aven and Jensen (1999).

For the purpose of comparing dependence properties of conditional distributions (given the values of some relevant parameters) with dependence properties of the predictive (unconditional) distributions, it can be useful for instance to take into account the above Theorem 3.43 and to combine the formulae (2.52), (2.53), and (2.59) of Chapter 2.

In order to specifically analyze dependence for the predictive distributions we take into account monotonicity properties of $\lambda_t^{(n-h)}(t_1, ..., t_h|\theta)$ and $\lambda_t^{(n)}(\theta)$ and stochastic comparisons for the posterior distributions of the parameter given different observed histories.

In this respect we note that, extending arguments presented in Remark 3.15, it is often possible to compare different posterior distributions in the sense of likelihood-ratio as will be discussed in the next subsection.

3.3.4 Likelihood-ratio comparisons between posterior distributions

Likelihood-ratio orderings among posterior distributions can be obtained by using orderings for the conditional distribution of data given the parameters (see e.g. Fahmi et al, 1982).

Statements of this type were presented in Section 3.1 of this chapter (Remark 3.15), for the special case when both the parameter Θ and the statistical observation X are scalar random quantities. Definitions of dependence presented so far put us in a position to extend such an analysis to the multivariate case.

For the following three results we consider the case when $(\mathbf{X}, \Theta)$ admits a joint density $f_{\mathbf{X},\Theta}(\mathbf{x},\theta)$, where $\Theta \equiv (\Theta_1, ..., \Theta_d)$ is interpreted as a parameter and $\mathbf{X} \equiv (X_1, ..., X_n)$ as a vector of observations.

$\pi_0(\theta)$ will denote the prior density of Θ; $f_{\mathbf{X}}(\mathbf{x}|\theta)$ will denote the conditional density of $\mathbf{X}$ given $(\Theta = \theta)$, and $\pi(\theta|\mathbf{x})$ will denote the posterior distribution of Θ, given a result $(\mathbf{X} = \mathbf{x})$.

When $\mathbf{X}$ is a vector of random lifetimes, the notation $\pi(\theta|\mathfrak{h}_t)$ will be used for the posterior distributions of Θ, given a history of the form (3.10).

We prefer to postpone proofs of the following results to the end of this subsection.

The following result is a slight modification of Theorem 3 in Fahmi et al. (1982). Here we assume that the set $L \equiv \{\theta \in \mathbb{R}^d | \pi_0(\theta) > 0\}$ is a *lattice* (see (3.6)).

Theorem 3.61. *Assume that*

a) for suitable functions $\psi:\mathbb{R}^n \to \mathbb{R}^m$, $h:\mathbb{R}^n \to \mathbb{R}_+$, and $g : \mathbb{R}^{m+d} \to \mathbb{R}_+$, $f_{\mathbf{X}}(\mathbf{x}|\theta)$ has the form

$$f_{\mathbf{X}}(\mathbf{x}|\theta) = h(\mathbf{x}) \cdot g(\psi(\mathbf{x}); \theta)$$

b) $g(q_1, ..., q_m; \theta_1, ..., \theta_d)$ and $\pi_0(\theta_1, ..., \theta_d)$ are MTP$_2$.
Then, for every $\mathbf{x}, \mathbf{x}'$ such that $\psi(\mathbf{x}) \geq \psi(\mathbf{x}')$,

$$\pi(\theta|\mathbf{x}) \geq_{lr} \pi(\theta|\mathbf{x}')$$

Note that the condition a) is equivalent to $\psi(\mathbf{x})$ being a *sufficient statistic* for Θ. Of course one can have in particular $m = n$ and $\psi(\mathbf{x}) \equiv \mathbf{x}$.

In our setting, i.e. in the case when $\mathbf{X}$ is a vector of lifetimes, one can need however to consider observed data of the form (3.10) rather than $\{\mathbf{X} = \mathbf{x}\}$.

This motivates the interest for the following two results. For simplicity we restrict attention to the case $d = 1$ (similar results might be proved also in the case $d > 1$, but, also this time, at the cost of more involved arguments and of additional MTP$_2$ conditions for related functions of $(\theta_1, ..., \theta_d)$).

Theorem 3.62. *Assume $f_{\mathbf{X}}(\mathbf{x}|\theta) \geq_{lr} f_{\mathbf{X}}(\mathbf{x}|\theta')$ for $\theta > \theta'$ (i.e. $\mathbf{X}$ is increasing in θ, in the $\geq_{lr}$ sense) and let $\mathfrak{h}'_t$ be more severe than $\mathfrak{h}_t$. Then*

$$\pi(\theta|\mathfrak{h}_t) \geq_{lr} \pi(\theta|\mathfrak{h}'_t).$$

In the next result we consider the special case of more severeness defined by (3.10) and (3.11) (a survival at t is replaced by a failure at t in the definition of $\mathfrak{h}'_t$). This allows us to achieve a result similar to Theorem 3.62 under a weaker condition. The interest of such a result lies in that it can be combined with Theorem 3.43, in order to prove properties of positive dependence.

Theorem 3.63. *Let* $\mathbf{X} \equiv (X_1, ..., X_n)$ *be increasing in* Θ *in the multivariate hazard rate ordering; then*

$$\pi(\theta|\mathfrak{h}_t) \geq_{lr} \pi(\theta|\mathfrak{h}'_t).$$

Remark 3.64. It can be of more general interest, in Bayesian analysis, to ascertain the existence of relations of stochastic orderings between two posterior distributions, corresponding to two different sets of data; such stochastic orderings can be used to obtain orderings between Bayes decisions, as we shall sketch in the last chapter. In particular, likelihood ratio orderings (and the related variation sign diminishing property) have a very important role in decision problems, as pointed out in the classical paper by Karlin and Rubin (1956).

We now turn to proving the results enunciated above.
Theorem 3.61

Proof. First we remark that, by Bayes' formula, one has

$$\pi(\theta|\mathbf{x}) = K(x)\pi_0(\theta)g(\psi(\mathbf{x}); \theta)$$

where $K(\mathbf{x}) = \frac{1}{\int_L \pi_0(\theta)g(\psi(\mathbf{x});\theta)d\theta}$. Taking into account a), b) and that $\psi(\mathbf{x}) \geq \psi(\mathbf{x}')$, we can obtain, for any $\theta, \theta' \in L$,

$$\pi(\theta \vee \theta'|\mathbf{x})\pi(\theta \wedge \theta'|\mathbf{x}') = K(\mathbf{x})K(\mathbf{x}')\pi_0(\theta \vee \theta')\pi_0(\theta \wedge \theta')\times$$

$$g(\psi(\mathbf{x}) \vee \psi(\mathbf{x}'); \theta \vee \theta')g(\psi(\mathbf{x}) \wedge \psi(\mathbf{x}'); \theta \wedge \theta') \geq$$

$$K(\mathbf{x})K(\mathbf{x}')\pi_0(\theta)\pi_0(\theta') \times g(\psi(\mathbf{x}); \theta)g(\psi(\mathbf{x}'); \theta') \ =$$

$$\pi(\theta|\mathbf{x})\pi(\theta'|\mathbf{x}').$$

$\square$

Theorem 3.62

Proof. Let

$$\mathfrak{h}_t \equiv \{\mathbf{X}_I = \mathbf{x}_I, \mathbf{X}_J > t\mathbf{e}_J, \mathbf{X}_{\widetilde{I \cup J}} > t\mathbf{e}_{\widetilde{I \cup J}}\}$$

$$\mathfrak{h}'_t \equiv \{\mathbf{X}_I = \mathbf{y}_I, \mathbf{X}_J = \mathbf{y}_J, \mathbf{X}_{\widetilde{I \cup J}} > t\mathbf{e}_{\widetilde{I \cup J}}\},$$

where $0 < y_i \le x_i < t$, for $i \in I, y_j \le t$, for $j \in J$.
We can write

$$\pi(\theta|\mathfrak{h}_t) \propto \pi(\theta|\mathbf{X}_{\widetilde{I \cup J}} > t\mathbf{e}_{\widetilde{I \cup J}}) f_{\mathbf{X}_I}(\mathbf{x}_I|\mathbf{X}_{\widetilde{I \cup J}} > t\mathbf{e}_{\widetilde{I \cup J}}; \theta) \times$$

$$\int_t^\infty \cdots \int_t^\infty f_{\mathbf{X}_J}(\xi_J|\mathbf{X}_I = x_I, \mathbf{X}_{\widetilde{I \cup J}} > t\mathbf{e}_{\widetilde{I \cup J}}; \theta) d\xi_J$$

$$\pi(\theta|\mathfrak{h}'_t) \propto \pi(\theta|\mathbf{X}_{\widetilde{I \cup J}} > t\mathbf{e}_{\widetilde{I \cup J}}) f_{\mathbf{X}_I}(\mathbf{y}_I|\mathbf{X}_{\widetilde{I \cup J}} > t\mathbf{e}_{\widetilde{I \cup J}}; \theta) \times$$

$$f_{\mathbf{X}_J}(\mathbf{y}_J|\mathbf{X}_I = \mathbf{y}_I, \mathbf{X}_{\widetilde{I \cup J}} > t\mathbf{e}_{\widetilde{I \cup J}}; \theta).$$

We should check that $\frac{\pi(\theta|\mathfrak{h}_t)}{\pi(\theta|\mathfrak{h}'_t)}$ is a non-decreasing function of θ. Now

$$\frac{\pi(\theta|\mathfrak{h}_t)}{\pi(\theta|\mathfrak{h}'_t)} = \int_t^\infty \cdots \int_t^\infty z(\xi_J; \theta, \mathbf{x}_I, \mathbf{y}_I, \mathbf{y}_J) d\xi_J$$

where

$$z(\xi_J; \theta, \mathbf{x}_I, \mathbf{y}_I, \mathbf{y}_J) \equiv$$

$$\frac{f_{\mathbf{X}_I}(\mathbf{x}_I|\mathbf{X}_{\widetilde{I \cup J}} > t\mathbf{e}_{\widetilde{I \cup J}}; \theta) f_{\mathbf{X}_J}(\xi_J|\mathbf{X}_I = x_I, \mathbf{X}_{\widetilde{I \cup J}} > t\mathbf{e}_{\widetilde{I \cup J}}; \theta)}{f_{\mathbf{X}_I}(\mathbf{y}_I|\mathbf{X}_{\widetilde{I \cup J}} > t\mathbf{e}_{\widetilde{I \cup J}}; \theta) f_{\mathbf{X}_J}(\mathbf{y}_J|\mathbf{X}_I = \mathbf{y}_I, \mathbf{X}_{\widetilde{I \cup J}} > t\mathbf{e}_{\widetilde{I \cup J}}; \theta)}$$

$$= \frac{f_{\mathbf{X}_I, \mathbf{X}_J}(\mathbf{x}_I, \xi_J|\mathbf{X}_{\widetilde{I \cup J}} > t\mathbf{e}_{\widetilde{I \cup J}}; \theta)}{f_{\mathbf{X}_I, \mathbf{X}_J}(\mathbf{y}_I, \mathbf{y}_J|\mathbf{X}_{\widetilde{I \cup J}} > t\mathbf{e}_{\widetilde{I \cup J}}; \theta)}.$$

Since, for $\theta > \theta'$, $f_{\mathbf{X}}(\mathbf{x}|\theta) \ge_{lr} f_{\mathbf{X}}(\mathbf{x}|\theta')$ it also follows that

$$f_{\mathbf{X}}(\cdot|\mathbf{X}_{\widetilde{I \cup J}} > t\mathbf{e}; \theta) \ge_{lr} f_{\mathbf{X}}(\cdot|\mathbf{X}_{\widetilde{I \cup J}} > t\mathbf{e}; \theta'),$$

by Theorem 3.22. Then, by the fundamental property that $\ge_{lr}$ is maintained for marginal distributions, it is, for $\theta > \theta'$

$$f_{\mathbf{X}_I, \mathbf{X}_J}(\mathbf{x}_I, \xi_J|\mathbf{X}_{\widetilde{I \cup J}} > t\mathbf{e}_{\widetilde{I \cup J}}; \theta) \ge_{lr} f_{\mathbf{X}_I, \mathbf{X}_J}(\mathbf{x}_I, \xi_J|\mathbf{X}_{\widetilde{I \cup J}} > t\mathbf{e}_{\widetilde{I \cup J}}; \theta').$$

By taking into account Remark 3.21, we can conclude the proof by noting that $z(\xi_J; \theta, \mathbf{x}_I, \mathbf{y}_I, \mathbf{y}_J)$ must be a non-decreasing function of θ, since $\mathbf{x}_I \ge \mathbf{y}_I$, and $\xi_J \ge t\mathbf{e} \ge \mathbf{y}_J$. $\qquad \square$

Theorem 3.63

Proof. Note that, by definition of m.c.h.r., the likelihood functions respectively associated to the observations $\mathfrak{h}_t$ and $\mathfrak{h}'_t$ are such that $L_{\mathfrak{h}'_t}(\theta) = \lambda_{i|I}(t|\mathbf{x}_I;\theta)L_{\mathfrak{h}_t}(\theta)$.

X increasing in Θ in the multivariate hazard rate ordering in particular implies that $\lambda_{i|I}(t|\mathbf{x}_I;\theta)$ is a non-increasing function of θ.

Then $\frac{\pi(\theta|\,\mathfrak{h}'_t)}{\pi(\theta|\,\mathfrak{h}_t)}$ is non-increasing in θ. $\qquad\qquad\qquad\Box$

3.4 Some notions of aging

Notions of *aging* are introduced, roughly speaking, to compare conditional survival probabilities for residual lifetimes of the type (3.1), for different *ages* of some of the surviving individuals. Some of these notions will be recalled in this section.

Our main purpose is to highlight part of the motivations for introducing notions of multivariate aging, that we shall call "Bayesian", in the analysis of exchangeable lifetimes; Bayesian notions will be studied in the next chapter.

Henceforth, we first recall some well-known concepts of one-dimensional aging; we also see the role of notions of univariate stochastic orderings in characterizing some of these concepts.

Later, we shall see how multivariate concepts of orderings, introduced in Section 3.2, are used to formulate some definitions of aging for vectors of random lifetimes. In the last subsection a short discussion will be presented about the range of applicability of such notions, in the frame of exchangeable lifetimes.

3.4.1 One-dimensional notions of aging

Let T be a lifetime with survival function $\overline{F}(t)$ and let

$$\overline{F}_s(t) = \frac{\overline{F}(t+s)}{\overline{F}(s)} = P\{T - s > t|T > s\}$$

denote the survival function of $T - s$, the residual lifetime at age s, conditional on survival at s.

The following definitions are well known:

The distribution of T is IFR (increasing failure rate) if

$$\overline{F}_s(t) \text{ is non-increasing in } s, \forall t > 0 \qquad (3.15)$$

or, equivalently,

$$\overline{F} \text{ is log-concave}$$

The distribution of T is DFR (decreasing failure rate) if

$$\overline{F}_s(t) \text{ is non-decreasing in } s, \forall t > 0 \qquad (3.16)$$

or, equivalently,

$$\overline{F} \text{ is log-convex}$$

The distribution of T is IFRA (increasing failure rate in average) if

$$\frac{-\log \overline{F}(t)}{t} \text{ is non-decreasing} \qquad (3.17)$$

The distribution of T is NBU new better than used) if

$$\overline{F}_s(t) \leq \overline{F}(t) \qquad (3.18)$$

The distribution of T is NWU (new worse than used) if

$$\overline{F}_s(t) \geq \overline{F}(t) \qquad (3.19)$$

The properties (3.15), (3.17), and (3.18) define notions of (one-dimensional) positive aging. Of course (3.15) is a special case of (3.17) which, in its turn, is a special case of (3.18); (3.16) is a special case of (3.19).

If the distribution admits a density function f, and then a failure rate function $r(t)$, properties (3.15), (3.16) are equivalent to $r(t)$ being non-decreasing and non-increasing, respectively.

Fix now a pair $s'' > s' > 0$ and consider the conditions

a) $\overline{F}_{s'}(t) \geq_{st} \overline{F}_{s''}(t), \forall t > 0$ and   b) $\overline{F}_{s'}(t) \geq_{hr} \overline{F}_{s''}(t), \forall t > 0$

a) is equivalent to

$$\text{a')} \quad \frac{\overline{F}(s'' + t)}{\overline{F}(s' + t)} \leq \frac{\overline{F}(s'')}{\overline{F}(s')} \ , \forall t > 0$$

while
b) is equivalent to

$$\text{b')} \quad \frac{\overline{F}(s'' + t)}{\overline{F}(s' + t)} \text{ decreasing in } t.$$

Obviously b') is a stronger condition than a').

However the condition that a') hold for any pair $0 \leq s' < s''$ is easily seen to be equivalent to the condition that b') hold for any pair $0 \leq s' < s''$ and both are equivalent to the IFR property of $\overline{F}$.

We can then conclude by noticing that the inequalities

$$\mathcal{L}[(X - t)|X > t] \geq_{hr} \mathcal{L}[(X - t')|X > t'] \ (0 \leq t < t'). \tag{3.20}$$

and

$$\mathcal{L}[(X - t)|X > t] \geq_{st} \mathcal{L}[(X - t')|X > t'] \ (0 \leq t < t'). \tag{3.21}$$

are both equivalent to the IFR property of $\overline{F}$.

This argument suggests, in the case of a lifetime T with a probability density function f, to consider the following stronger concept of positive aging:

$$\mathcal{L}[(T - t)|T > t\,] \geq_{lr} \mathcal{L}[(T - t')|T > t'], \text{ for } 0 \leq t < t'. \tag{3.22}$$

The comparison (3.22) is equivalent to the condition that the distribution of T is PF_2 (Polya frequency of order 2) namely that the density f is a log-concave function on $[0, \infty)$.

We note also that the notions of NBU and NWU can be rephrased in terms of stochastic orderings; indeed T is NBU (NWU) if and only if

$$\mathcal{L}(X) \geq_{st} \ (\leq_{st}) \ \mathcal{L}[(X - t)|X > t], \text{ for } 0 < t. \tag{3.23}$$

Let us now think of the case when the survival function of a lifetime T is assessed conditionally on the value of some parameter Θ; denoting the conditional survival function by $\overline{G}(\cdot|\theta)$ and assuming that Θ is a random quantity, taking values in a set L with a prior density $\pi_0(\theta)$, we obtain (recall Example 2.9, Chapter 2) that the "predictive" survival function of T is

$$\overline{F}(s) = \int_L \overline{G}(s|\theta) \pi_0(\theta) \, d\theta \tag{3.24}$$

and we can write the predictive survival function of the residual lifetime at age s in the form

$$\overline{F}_s(t) = \int_L \overline{G}_s(t|\theta) \pi_s(\theta|T > s) \, d\theta$$

where

$$\overline{G}_s(t) = \frac{\overline{G}(t + s|\theta)}{\overline{G}(s|\theta)}, \ \pi_s(\theta|T > s) = \frac{\overline{G}(s|\theta) \pi_0(\theta) \, d\theta}{\int_L \overline{G}(s|\theta) \pi_0(\theta) \, d\theta}.$$

A central point is the comparison between aging properties of the conditional survival functions $\overline{G}(t|\theta)$, $\theta \in L$, and of the predictive survival function $\overline{F}(t)$. The following facts are well known and can have a big impact in the applications

Proposition 3.65. *(see Barlow and Proschan, 1975). If $\overline{G}(t|\theta)$ is DFR $\forall \theta \in L$, then $\overline{F}(t)$ is DFR.*

Remark 3.66. Being $\overline{G}(t|\theta)$ IFR $\forall \theta \in L$, on the contrary, does not imply that $\overline{F}(t)$ is IFR.

This fact provides an explanation of an apparent paradox that often arises in engineering practice (see e.g. Proschan, 1963 and Barlow, 1985). It formally describes the following case:

we start with a "pessimistic" state of information about Θ (i.e. we expect that Θ is such as to prevent a large value of the lifetime T); pessimism is contradicted by progressive observation of survival, so that we may become more and more optimistic, even if, per se, increasing age deteriorated the unit.

These facts can be e.g. illustrated by means of the example considered in Exercise 2.64 where the conditional failure rates are $r(t|\theta) = \theta t^\varepsilon$ and the predictive failure rate is $r(t) = t^\epsilon/(\beta + t^{1+\epsilon})$. For $\epsilon > 0$, $r(t|\theta)$ are increasing for all $t > 0$, $\theta > 0$ while $r(t)$ is increasing in a neighborhood of the origin and decreasing for t large enough. For $-1 < \epsilon < 0$, $r(t|\theta)$ are decreasing for all $t > 0$, $\theta > 0$ and $r(t)$ is decreasing as well.

The following remarks are relevant for the arguments that will be developed in the next chapter.

Remark 3.67. Each of the notions IFR, DFR, NBU, NWU can be also looked on as an appropriate inequality for the corresponding joint distribution of n i.i.d. lifetimes. Let us focus attention, for instance, on the notion of IFR and consider lifetimes $T_1, ..., T_n$ which are i.i.d with a common survival function $\overline{G}(\cdot)$. $\overline{G}(\cdot)$ being IFR is equivalent to requiring

$$P\{T_1 > s_1 + \tau|D\} \geq P\{T_2 > s_2 + \tau|D\} \tag{3.25}$$

where $D \equiv \{T_1 > s_1, T_2 > s_2, ..., T_k > s_k, T_{k+1} = t_{k+1}, ..., T_n = t_n\}$ and whenever $s_1 < s_2$.

In the more general case when $T_1, ..., T_n$ are exchangeable, the IFR property of their marginal one-dimensional distribution is not anymore equivalent to the inequality (3.25). However we can notice that, in the case of stochastic dependence, we are, in general, more interested in the validity of (3.25) rather than in the IFR property of the marginal.

We can say that, when analyzing a vector of several exchangeable lifetimes, the marginal IFR property is really relevant only in the case of independence.

Remark 3.68. Proposition 3.65 and Remark 3.66, respectively, can be rephrased in the following equivalent ways:

if $\overline{G}(t|\theta)$ is log-convex $\forall \theta \in L$, then $\overline{F}(t)$ in (3.24) is log-convex for any initial density π_0

$\overline{G}(t|\theta)$ being log-concave $\forall \theta \in L$ does not imply that $\overline{F}(t)$ in (3.24) is log-concave.

The theme of aging properties of one-dimensional mixture distributions is of great interest in the field of reliability and survival analysis; there is then a very rich and still increasing literature on this topic. Many authors focused attention on relations between the behavior of failure rate of mixtures and aging properties of single components of the mixture; this has both an interest in the theory and in the applied statistical analysis. We refer in particular to Manton, Stallard and Vaupel (1979), Vaupel and Yashin (1985), Hougaard (1986), Cohen (1986), Gurland and Sethuraman (1995), among many others; see also the review paper by Shaked and Spizzichino (2001).

In such a frame it is particularly important, however, to keep in mind that, under suitable conditions, a mixture of IFR distribution can still be IFR itself; sufficient conditions in this sense are shown by Lynch (1999).

In the field of reliability, the study of aging properties of mixtures is relevant in the analysis of burn-in problems (see e.g. the review paper by Block and Savits, 1997). In this respect, the discussion in the following example is of interest for our purposes.

Example 3.69. (The burn-in problem). Let U be a unit to be put into operation; T denotes its lifetime and its survival function is $\overline{F}(s)$.

We get a reward K under the condition that, in its operative life, it survives a mission time τ; but we incur a loss C if the operative life is smaller than τ.

We can decide not to put U into operation (i.e. to discard it) and this causes a cost c $(0 < c < K < C)$; if we decide that it is worth it to put U into operation, we may prefer to deliver it to operations after a burn-in period. If we choose σ as the duration of the burn-in period, then the length of the operative life (under the condition $T > \sigma$, of course) will be the residual lifetime $(T - \sigma)$ and its (conditional) survival function is

$$\overline{F}_\sigma(t) \equiv P\{T - \sigma > t | T > \sigma\} = \frac{\overline{F}(\sigma+t)}{\overline{F}(\sigma)}$$

Suppose, moreover, that there is no cost for conducting the burn-in test and that we incur the cost c if $T < \sigma$, namely if U fails during the test (i.e. this would have the same economical effect as the decision of discarding U).

The expected cost coming from the choice of a duration σ for the test would then be

$$\phi_\sigma = c \cdot P\{T < \sigma\} + C \cdot P\{\sigma < T < \sigma + \tau\} - K \cdot P\{T > \sigma + \tau\}$$
$$= c + (C - c)\overline{F}(\sigma) - (K + C)\overline{F}(\sigma + \tau)$$

and the optimal (i.e. Bayes) duration σ^* is such to minimize ϕ_σ (see details in Chapter 5).

We can have three cases
a)

$$(C - c)\overline{F}(\sigma) - (K + C)\overline{F}(\sigma + \tau) > 0, \forall \sigma > 0$$

b)

$$\inf_{\sigma \geq 0} (C - c)\overline{F}(\sigma) - (K + C)\overline{F}(\sigma + \tau) =$$

$$(C - c) - (K + C)\overline{F}(\tau) < 0$$

c) For some $\sigma > 0$, it is

$$(C - c)\overline{F}(\sigma) - (K + C)\overline{F}(\sigma + \tau) < \min\{(C - c) - (K + C)\overline{F}(\tau), 0\}$$

In case a) it is optimal to discard U;

in case b) it is optimal to deliver immediately U to operations;

in case c) we can find $\sigma^* > 0$ such that it is optimal to conduct a burn-in test of duration σ^* and then deliver U to operation if it survives the test.

Trivially c) can never apply if $\overline{F}(s)$ is NBU; it may apply in other cases, for instance when $\overline{F}(s)$ is NWU.

In particular it may be optimal to conduct a burn-in procedure when T is conditionally exponential given a parameter θ, i.e.

$$\overline{F}(s) = \int_0^\infty \exp\{-\theta s\}\pi_0(\theta)d\theta.$$

Reliability engineers may think that it is senseless to burn-in a unit with an exponential lifetime, so that the latter conclusion may appear contradictory.

The question is: "does it really make a difference if θ is known or unknown?"

Of course it makes a great difference in the Bayesian paradigm.

Indeed the latter prescribes that decisions are to be taken by minimizing expected loss, which is to be computed taking into account one's own utility function and prediction based on actual state of information (see Section 1 of Chapter 5).

This means the following:

Consider two individuals I_1 and I_2, both sharing the same utilities and convinced that the distribution of T is influenced by the factor Θ. However, according to I_1's initial state of information, the density of Θ is $\pi^{(1)}$ and the conditional survival function of T given $(\Theta = \theta)$ is $\overline{G}^{(1)}(s|\theta)$ while I_2 assesses $\pi^{(2)}$ and $\overline{G}^{(1)(2)}(s|\theta)$ respectively.

If, by chance, it happens that

$$\int_0^\infty \overline{G}^{(1)}(s|\theta)\pi^{(1)}(\theta)d\theta = \int_0^\infty \overline{G}^{(2)}(s|\theta)\pi^{(2)}(\theta)d\theta, \ \forall s > 0 \qquad (3.26)$$

then, in a Bayesian viewpoint, I_1 and I_2 should make the same decision as to the problem of considering a burn-in procedure, irrespective of their different opinions. Then unidimensional aging properties of $\overline{G}^{(i)}(s|\theta)$ have not a direct impact on the decision; only one-dimensional aging properties of the predictive survival functions $\int_0^\infty \overline{G}^{(i)}(s|\theta)\pi^{(i)}(\theta)d\theta$ are relevant.

3.4.2 Dynamic multivariate notions of aging

Here we discuss some formulations of the concept of "aging" for a vector of lifetimes. In the literature, several different concepts have been proposed to extend the univariate concept of IFR distribution to the multivariate case; we shall report some of those concepts here.

We saw above that a lifetime X is IFR if and only if, for all $0 \leq t < t'$ with t' such that $\overline{F}(t') > 0$, the inequality (3.20) is true.

This suggested a definition of $MIFR$ (multivariate increasing failure rate) presented in Shaked and Shanthikumar (1991), which we report here in the following, slightly modified, form.

Definition 3.70. A vector of lifetimes $\mathbf{X} \equiv (X_1, ..., X_n)$ is $MIFR$ when, for $0 \leq t < t', I \cap J = \emptyset, \mathbf{x}'_I \leq \mathbf{x}_I \leq t\mathbf{e}, t\mathbf{e} < \mathbf{x}'_J \leq t'\mathbf{e}$, and histories $\mathfrak{h}_t, \mathfrak{h}'_{t'}$ of the form

$$\mathfrak{h}_t \equiv \{\mathbf{X}_I = \mathbf{x}_I; \mathbf{X}_{\bar{I}} > t\mathbf{e}\}, \mathfrak{h}'_{t'} \equiv \{\mathbf{X}_I = \mathbf{x}'_I, \mathbf{X}_J = \mathbf{x}_J; \mathbf{X}_{\overline{I \cap J}} > t'\mathbf{e}\} \qquad (3.27)$$

one has

$$\mathcal{L}[(\mathbf{X}_{\overline{I \cup J}} - t\mathbf{e})|\mathfrak{h}_t] \geq_{hr} \mathcal{L}[(\mathbf{X}_{\overline{I \cup J}} - t'\mathbf{e})^+|\mathfrak{h}'_{t'}] \qquad (3.28)$$

Note that, in this definition, the histories $\mathfrak{h}_t$ and $\mathfrak{h}'_{t'}$ coincide over the interval $[0, t]$.

We saw above that, in the one-dimensional case, the inequality (3.20) is equivalent to the apparently weaker condition (3.21).

Of course that cannot be true in general for the multivariate case. This fact then leads to the definition of a weaker concept of $MIFR$, which can be seen as a special case of the notion of $\mathcal{F}_t$-$MIFR$, analyzed by Arjas (1981a).

Definition 3.71. A vector of lifetimes $\mathbf{X} \equiv (X_1, ..., X_n)$ is $st - MIFR$ if, for $0 \leq t < t'$ and histories $\mathfrak{h}_t$, $\mathfrak{h}'_{t'}$ of the form (3.27), one has

$$\mathcal{L}[(\mathbf{X}_{\overline{I \cup J}} - t\mathbf{e})|\mathfrak{h}_t] \geq_{st} \mathcal{L}[(\mathbf{X}_{\overline{I \cup J}} - t'\mathbf{e})|\mathfrak{h}'_{t'}] \tag{3.29}$$

On the other side, the following definition is a natural multivariate analogous of the PF_2 property.

Definition 3.72. (Shaked and Shanthikumar, 1991) $\mathbf{X}$ is MPF_2 if and only if, for $0 \leq t < t'$ and histories $\mathfrak{h}_t$, $\mathfrak{h}'_{t'}$ one has

$$\mathcal{L}[(\mathbf{X}_{\overline{I \cup J}} - t\mathbf{e})|\mathfrak{h}_t] \geq_{lr} \mathcal{L}[(\mathbf{X}_{\overline{I \cup J}} - t'\mathbf{e})|\mathfrak{h}'_{t'}] \tag{3.30}$$

whenever $\mathfrak{h}_t$, $\mathfrak{h}'_{t'}$ are of the form (3.27).

The definitions recalled above are conditional and of dynamic type, in that they are formulated in terms of the comparison between conditional laws of residual lifetimes at two different instants t, t', given histories $\mathfrak{h}_t$, $\mathfrak{h}'_{t'}$, which coincide over the interval $[0, t]$.

A definition of MIFR which, on the contrary, is not of this type was given in Savits (1985) (see also Shaked and Shanthikumar, 1978, for some related characterization).

For other definitions, which do not directly involve conditional distributions of residual lifetimes given failure and survival data, one can see e.g. Harris (1970) and Marshall (1975).

Now we highlight some aspects of the above dynamic definitions.

First of all we note that, by taking into account the chain of implications existing among multivariate stochastic orders (Theorem 3.28), one gets

Theorem 3.73.

$$\mathbf{X} \; MPF_2 \Rightarrow \mathbf{X} \; MIFR \Rightarrow \mathbf{X} \; st\text{-}MIFR$$

Remark 3.74. A common feature of the above definitions is that they imply different types of positive dependence properties. More precisely, the following results hold:

  i) $\mathbf{X} \; MPF_2 \Rightarrow \mathbf{X} \; MTP_2$ (Shaked and Shanthikumar, 1991, Remark 6)
  ii) $\mathbf{X} \; MIFR \Rightarrow \mathbf{X} \; HIF$ (Shaked and Shanthikumar, 1991, Remark 5)
  iii) $\mathbf{X} \; st - MIFR \Rightarrow WBF$ (Norros, 1985).

Example 3.75. (Ross models). Consider a (non-necessarily exchangeable) Ross model where $(X_1, ..., X_n)$ has an absolutely continuous joint distribution with m.c.h.r. functions such that

$$\lambda_{i|I}(t|x_1, ..., x_h) = \lambda_i(I)$$

(with $h = |I|$). In such a case the $MIFR$ property amounts to requiring that $\lambda_i(I) \leq \lambda_i(J)$, whenever $I \subset J$ (see also Ross, 1984; Norros, 1985; Shaked and

Shanthikumar, 1991). Then, for such a model, $\mathbf{X}$ $MIFR$ is equivalent to $\mathbf{X}$ HIF.

Remark 3.76. Let $X_1, ..., X_n$ be independent lifetimes with one-dimensional failure rates $r_i(\cdot)$ and densities $f_i(\cdot)$. Then
$\mathbf{X} \equiv (X_1, ..., X_n)$ is $st\text{-}MIFR$ if and only if $r_i(\cdot)$ is an increasing function $(i = 1, ..., n)$, which is also equivalent to $\mathbf{X}$ being $MIFR$.
$\mathbf{X}$ is MPF_2 if and only if $f_i(\cdot)$ is a PF_2 function $(i = 1, ..., n)$.

More generally, the one-dimensional marginal distributions of a random vector with the MIFR property is IFR.

Many other multivariate extensions of the property of IFR require that the marginal distributions are univariate IFR.

An important remark is that any plausibly defined property of aging is subject to be altered or destroyed when we modify the actual flow of information.

Then we should clarify, in the definition of a notion of aging, which is the state of information upon which we condition, when considering the distribution of residual lifetimes of still surviving individuals.

Here we considered the special case when, at any time t, conditioning is made with respect to the *internal history*, i.e. to a state of information of the type

$$\mathfrak{h}_t \equiv \{\mathbf{X}_I = \mathbf{x}_I; \mathbf{X}_{\bar{I}} > t e\}.$$

For a much more general treatment and detailed discussion see Arjas (1981a), Arjas (1981b), Arjas and Norros (1991).

Many other properties and aspects of Definitions 3.70, 3.71, 3.72 were proved and discussed in more general settings in Arjas (1981a), Arjas (1981b), Arjas and Norros (1984), Norros (1985), Arjas and Norros (1991), Shaked and Shanthikumar (1991), Shaked and Shanthikumar (1994) and references therein.

3.4.3   The case of exchangeable lifetimes

Concerning the above definitions of multivariate aging, we want to discuss here some aspects related to the kind of interdependence described by the notion of exchangeability.

In other words, we consider the case of similar individuals, which are embedded in the same environment, which undergo the same situation of stress, and which are subjected to the same shocks, so that the probability model which describes our uncertainty about the corresponding lifetimes is exchangeable (we are thinking of the cases when the lifetimes are not independent).

Even if we admit that any single individual deteriorates in time, we cannot expect in general that the vector of lifetimes satisfies any of the dynamic conditions of positive aging, defined in the previous subsection.

In fact the latter conditions require that the lifetimes are positively dependent as a consequence of the results cited in Remark 3.74.

We can then exclude in particular the validity of MIFR, in the cases of negative dependence.

Example 3.77. (Negative dependence) Let $\mathbf{T} \equiv (T_1, T_2)$ have an exchangeable joint distribution characterized by the set of m.c.h.r. functions

$$\lambda^{(2)}(t) = r_2(t), \lambda_t^{(1)}(t_1) = r_1(t) \qquad (3.31)$$

If $r_2(t) \geq r_1(t)$, this is a case of negative dependence and, accordingly, $\mathbf{T}$ is not WBF. Then $\mathbf{T}$ cannot be $MIFR$, even if $r_1(t)$ and $r_2(t)$ are increasing functions.

On the other hand, in the case of exchangeability, positive dependence is often concomitant with situations of uncertainty about some unobserved quantity.

In its turn, this implies that the predictive distributions are of mixture-type and we cannot expect that conditions of positive aging generally hold.

We now spell out these arguments in more precise detail, by focusing attention on the notion of MIFR, for exchangeable lifetimes.

Let us then consider an exchangeable vector $\mathbf{T} \equiv (T_1, ..., T_n)$ with joint distribution admitting a density, and then m.c.h.r. functions. The m.c.h.r. functions will be, as usual, denoted by $\lambda_t^{(n-h)}(t_1, ..., t_h)$ and $\lambda_t^{(n)}$.

We can notice that, as a necessary condition for MIFR, we must have in particular

$$\lambda_t^{(n)} \leq \lambda_{t'}^{(n)}, \ \lambda_t^{(n-h)}(t_1, ..., t_h) \leq \lambda_{t'}^{(n-h)}(t_1, ..., t_h) \qquad (3.32)$$

where $t_1, ..., t_h < t$.

Let us focus attention on those cases of exchangeable, positive dependence when, conditional on a given parameter Θ, $T_1, ..., T_n$ are exchangeable and the corresponding conditional distributions are characterized by a system of m.c.h.r. functions $\lambda_t^{(n-h)}(t_1, ..., t_h | \theta)$ and $\lambda_t^{(n)}(\theta)$. Let L denote the set of possible values of Θ and let $\pi(\theta | \mathfrak{h}_t)$ denote the conditional density, given the observation of a dynamic history $\mathfrak{h}_t$.

Recall now the Theorem 2.43. The conditions (3.32) become, respectively,

$$\int_L \lambda_t^{(n)}(\theta) \pi(\theta | T_{(1)} > t) d\theta \ \leq \ \int_L \lambda_{t'}^{(n)}(\theta) \pi(\theta | T_{(1)} > t') d\theta \qquad (3.33)$$

and

$$\int_L \lambda_t^{(n-h)}(t_1, ..., t_h | \theta) \pi(\theta | \mathfrak{h}_t) d\theta \ \leq \ \int_L \lambda_{t'}^{(n-h)}(t_1, ..., t_h | \theta) \pi(\theta | \mathfrak{h}'_{t'}) d\theta \qquad (3.34)$$

where $0 \leq t_1 < ... < t_h \leq t < t'$, and

$$\mathfrak{h}_t \equiv \{T_{(1)} = t_1, ..., T_{(h)} = t_h, T_{(h+1)} > t\}$$

$$\mathfrak{h}'_{t'} \equiv \{T_{(1)} = t_1, ..., T_{(h)} = t_h, T_{(h+1)} > t'\}.$$

Specifically, when $T_1, ..., T_n$ are conditionally i.i.d. given Θ, denoted by $r(t|\theta)$ the conditional hazard rate functions, then in particular the conditions (3.33) and (3.34) become

$$\int_L r(t|\theta)\pi(\theta|T_{(1)} > t)d\theta \leq \int_L r(t'|\theta)\pi(\theta|T_{(1)} > t')d\theta$$

and

$$\int_L r(t|\theta)\pi(\theta|\mathfrak{h}_t)d\theta \leq \int_L r(t'|\theta)\pi(\theta|\mathfrak{h}'_{t'})d\theta$$

Remark 3.78. (3.33) and (3.34) constitute a very strong set of conditions, which can seldom be expected to hold, even if some strong positive dependence property is assumed for $\mathbf{T} \equiv (T_1, ..., T_n)$ and if $\lambda_t^{(n)}(\theta)$ and $\lambda_t^{(n-h)}(t_1, ..., t_h|\theta)$ are increasing functions of t, $\forall \theta \in L$.

Suppose for instance that Θ is a scalar quantity and $\lambda_t^{(n)}(\theta)$ is an increasing function of θ and t. Then, by a fundamental property of the one-dimensional $\leq_{st}$ ordering, (3.33) would be guaranteed by the condition

$$\mathcal{L}(\Theta|T_{(1)} > t) \leq_{st} \mathcal{L}(\Theta|T_{(1)} > t'), \text{ for } t < t'$$

The latter inequality is however just in contrast with the assumption that $\lambda_t^{(n)}(\theta)$ is increasing in $\theta, \forall t > 0$ (see Remark 3.15). Similar arguments can be repeated for (3.34), by using Theorem 3.62.

The points above are illustrated by the following examples. In the first example we analyze the (limiting) case of Schur-constant densities.

Example 3.79. (Schur-constant densities). Let $\mathbf{T}$ have a joint density function $f(\mathbf{t}) = \phi(\sum_{i=1}^n t_i)$, for a suitable function ϕ. Then, as we saw in Example 3.45, $\mathbf{T}$ is WBF if and only if ϕ is log-convex, i.e. positive dependence goes together with negative aging of the one-dimensional marginal distribution. We can see that $\mathbf{T}$ is $MIFR$ if and only if $T_1, ..., T_n$ are independent, exponentially distributed (i.e. $\phi(t) = \theta^n \exp\{-\theta \cdot t\}$, for some $\theta > 0$). This fact can also be seen by recalling that, in the absolutely continuous Schur-constant case,

$$\lambda_t^{(n-h)}(t_1, ..., t_h) = \widehat{\lambda}(h, y) \equiv -\overline{\Phi}^{(h+1)}(y)/\overline{\Phi}^{(h)}(y)$$

where $\overline{\Phi}^{(n)} = (-1)^n \phi, y = \sum_{i=1}^n t_i + (n - h)t$ and noticing that the condition of $MIFR$ in particular implies that $\widehat{\lambda}(h, y)$ is a non-decreasing function of y.

Example 3.80. The arguments in the example above can be in some sense extended to "time-transformed exponential models" defined by the condition

$$\overline{F}^{(n)}(\mathbf{t}) = \overline{\Phi}(\sum_{i=1}^{n} R(t_i)).$$

For these models, cases of absolute continuity and positive dependence are of the type

$$\overline{F}^{(n)}(\mathbf{t}) = \int_0^{+\infty} \exp\{-\theta \cdot \sum_{i=1}^{n} R(t_i)\} d\Pi_0(\theta)$$

with R an increasing, differentiable function. Letting $\rho(t) = \frac{d}{dt}R(t)$, we have that the functions $\lambda_t^{(n)}$, $\lambda_t^{(n-h)}(t_1, ..., t_h)$ are of the form

$$\lambda_t^{(n)} = \rho(t)\mathbb{E}\left[\Theta|T_{(1)} > t\right]$$
$$\lambda_t^{(n-h)}(t_1, ..., t_h) = \rho(t)\mathbb{E}\left[\Theta|T_{(1)} = t_1, ..., T_{(h)} = t_h, T_{(h+1)} > t\right]$$

(see Example 2.40). We can remark now that

$$\mathbb{E}\left[\Theta|T_{(1)} > t\right] \text{ and } \mathbb{E}\left[\Theta|T_{(1)} = t_1, ..., T_{(h)} = t_h, T_{(h+1)} > t\right]$$

are non-increasing functions of t; then we cannot expect that the conditions (3.32) generally hold, even if ρ is increasing, i.e. even if $T_1, ..., T_n$ are conditionally i.i.d IFR.

The following example more generally shows that a vector $\mathbf{T}$ being MIFR conditionally on $\{\Theta = \theta\}$, $\forall \theta \in L$, does not imply that it is st-MIFR in the unconditional distribution.

Example 3.81. (Conditional exchangeable Ross models). Let Θ be a nonnegative random quantity seen as an unobservable parameter and consider $\mathbf{T} \equiv (T_1, T_2)$ which, conditionally on $\{\Theta = \theta\}(\theta > 0)$, is an exchangeable Ross model with

$$\lambda_i(\emptyset) = \theta, \lambda_i(\{j\}) = 2\theta, i, j = 1, 2; i \neq j.$$

Then $\mathbf{T}$ is *MIFR* in the conditional distribution, given $\{\Theta = \theta\}$, $\forall \theta > 0$.

Let π_0 be the prior density of Θ and fix now $t' > t_1 > t \geq 0$; we can write

$$P\{T_2 > t' + s|T_1 = t_1, T_2 > t'\} =$$
$$\int_0^\infty P\{T_2 > t' + s|T_1 = t_1, T_2 > t'; \theta\} \cdot \pi(\theta|T_1 = t_1, T_2 > t')d\theta$$
$$P\{T_2 > t + s|T_1 > t, T_2 > t\} =$$
$$\int_0^\infty P\{T_2 > t + s|T_1 > t, T_2 > t; \theta\} \cdot \pi(\theta|T_1 > t, T_2 > t')d\theta.$$

$P\{T_2 > t' + s | T_1 = t_1, T_2 > t'; \theta\}$ and $P\{T_2 > t + s | T_1 > t, T_2 > t; \theta\}$ can be easily computed by taking into account the special form of $\lambda_i(\emptyset)$, $\lambda_i(\{j\})$.

$P\{T_2 > t' + s | T_1 = t_1, T_2 > t'\}$ and $P\{T_2 > t + s | T_1 > t, T_2 > t\}$ can be explicitly computed by taking, e.g.,

$$\pi_0(\theta) \propto \theta^{\alpha-1} \exp\{-\beta\theta\}, \theta > 0$$

(see Scarsini and Spizzichino, 1999, for details).

One can choose t' large enough and s small enough, in order to have

$$P\{T_2 > t' + s | T_1 = t_1, T_2 > t'\} > P\{T_2 > t + s | T_1 > t, T_2 > t\}.$$

Then $\mathbf{T}$ is not $st - MIFR$ in the unconditional distribution.

In conclusion, let us come back to the case of "exchangeable", deteriorating, individuals, as considered at the beginning of this subsection.

The arguments above point out the following: any notion of multivariate aging which requires, at a same time, both positive dependence and, at least, some positive aging property in the sense of inequalities (3.32) cannot be expected generally to hold for vectors of exchangeable lifetimes.

We can then wonder which type of inequalities could, in the case of exchangeability, be associated with a judgment of wear-out.

This problem gives us part of the motivation for the study of the notions which will be considered in the following chapter.

3.5   Exercises

Exercise 3.82. Check that $\mathbf{T} \geq_{lr} \widehat{\mathbf{T}}$ for the exchangeable Ross models characterized by the systems of m.c.h.r. functions

$$\lambda_t^{(n-h)}(t_1, ..., t_h) = \frac{\theta}{n-h}, \widehat{\lambda}_t^{(n-h)}(t_1, ..., t_h) = \frac{\widehat{\theta}}{n-h}$$

with $0 < \theta \leq \widehat{\theta}$. Using this result we see, in particular, that $\mathbf{T}$ is MTP_2.

Hint: Recall that, if for given $\theta > 0$, $\varphi_h(t) = \frac{\theta}{n-h}$, then the corresponding joint density function is

$$f^{(n)}(t_1, ..., t_n) = \frac{\theta^n}{n!} \exp\{-\theta t_{(n)}\}.$$

Exercise 3.83. Consider, more generally, two vectors of lifetimes $\mathbf{T}$ and $\mathbf{T}'$ corresponding to models characterized by m.c.h.r. functions of the form

$$\lambda_t^{(n-h)}(t_1, ..., t_h) = \varphi_h(t), \mu_t^{(n-h)}(t_1, ..., t_h) = \rho\varphi_h(t),$$

respectively, where ρ is a positive quantity. Show that, if $\mathbf{T}$ and $\mathbf{T}'$ are HIF and $\rho > 1$, then $\mathbf{T} \geq_{hr} \mathbf{T}'$.

The following result is well known. It provides an important characterization of the condition $X \leq_{st} Y$. Let X, Y be two n-dimensional random vectors. The condition $X \leq_{st} Y$ holds if and only if we can find a probability space $(\Omega, \mathcal{F}, P)$ and two n-dimensional random vectors $\widehat{X}, \widehat{Y} \colon \Omega \to R^n$ such that

$$\widehat{X}(\omega) \leq \widehat{Y}(\omega), \forall \omega \in \Omega; \widehat{X} =_{st} X, \widehat{Y} =_{st} Y.$$

Exercise 3.84. By using the result above, check the inequality in Example 3.34.

Exercise 3.85. Consider a proportional hazard model $\overline{F}_{R,\pi_0}$. Check directly the validity of the MTP_2 property, when the initial density π_0 is of the gamma type.

Exercise 3.86. For the disruption model considered in Example 3.47, spell out arguments, similar to those in the Remark 3.15, to check the WBF property under the condition $r_1(t) \leq r_2(t)$.

Exercise 3.87. Consider the heterogeneous population in Example 2.20. There we have $\Theta \equiv (K_1, ..., K_n)$ where K_j is a binary random variable explaining the "subpopulation" which the individual U_j belongs to; for the individuals' lifetimes $T_1, ..., T_n$ we assumed that they are conditionally independent given Θ and that

$$P\{T_j > t | \Theta = (k_1, ..., k_n)\} = \overline{G}_j(t)$$

where $k_1, ..., k_n \in \{0, 1\}$ and $\overline{G}_0, \overline{G}_1$ are two given (one-dimensional) survival functions.

Show that $T \equiv (T_1, ..., T_n)$ is PUOD if Θ is positively correlated and $\overline{G}_0(t) \leq \overline{G}_1(t)$ (or $\overline{G}_0(t) \geq \overline{G}_1(t)$) for any $t > 0$.

Exercise 3.88. Under the same setting and notation of the exercise above, show that T is WBF if $\overline{G}_0 \leq_{hr} \overline{G}_1$ and $P\{C_1 = 1 | \mathfrak{h}_t\} \geq P\{C_1 = 1 | \mathfrak{h}'_t\}$ where $\mathfrak{h}_t$, $\mathfrak{h}'_t$ are two histories as in (3.10) and (3.11).

Hint: note that, for a history $\mathfrak{h}_t$,

$$P\{T_1 > t + s | \mathfrak{h}_t\} =$$
$$\frac{G_0(t + s)}{G_0(t)} P\{C_1 = 0 | \mathfrak{h}_t\} + \frac{G_1(t + s)}{G_1(t)} P\{C_1 = 1 | \mathfrak{h}_t\}.$$

Exercise 3.89. (Schur-constant survival functions). Let $T \equiv (T_1, ..., T_n)$ have a Schur-constant joint survival function: $\overline{F}(t_1, ..., t_n) = \overline{\Phi}(\sum_{i=1}^{n} t_i)$, for a suitable non-increasing function $\overline{\Phi}$. Show that
$\overline{F} \in \Delta_{PUOD}$ if and only if T_i is NWU
and
$\overline{F} \in \Delta_{NUOD}$ if and only if T_i is NBU

Hint: Remember that, in this case, the one-dimensional survival function of any life-time T_i is given by

$$\overline{F}^{(1)}(t) = P\{T_1 > t, T_2 > 0, ..., T_n > 0\} = \overline{\Phi}(t).$$

Exercise 3.90. Show that Schur-constant joint survival functions with the MTP_2 property have DFR one-dimensional marginals (this then happens for conditionally independent, exponential lifetimes).

Exercise 3.91. Consider the special case of the Schur-constant joint survival function with

$$\overline{\Phi}(t) = \frac{1}{s}[s - t]_+^{n-1}$$

($[a]_+ = a$ if $a \geq 0$ and $[a]_+ = 0$ if $a < 0$).

Show that $(T_1, ..., T_{n-1})$ satisfies the conditions N.

Hint: This distribution corresponds to

$$\overline{F}(t_1, ..., t_n) = P\{X_1 > t_1, ..., X_n > t_n \mid \sum_{i=1}^{n} X_i = s\}$$

with $X_1, ..., X_n$ i.i.d. exponentially distributed ("uniform distribution over the simplex $\sum_{i=1}^{n} T_i = s$"). It is also $\overline{F} \in \Delta_{NA}$ (Joag-Dev and Proschan, 1983, Theorem 2.8).

Exercise 3.92. Check that $\mathbf{T} \equiv (T_1, T_2)$ in Example 3.31, where

$$\lambda_t^{(2)} = \frac{\theta + a}{2}, \lambda_t^{(1)}(t_1) = \theta,$$

is RR_2 for $a > \theta$.

Hint: Recall the Formula (2.58).

Exercise 3.93. Consider again the case of the heterogeneous population as in the above Exercise 3.87. Show that, under the condition $\overline{G}_0(t) \leq \overline{G}_1(t)$ (or $\overline{G}_0(t) \geq \overline{G}_1(t)$) for any $t > 0$, $\mathbf{T}$ is NUOD if Θ is negatively correlated.

Exercise 3.94. For the same case as above, and comparing also with Exercise 3.88, deduce that the NWU of the one-dimension marginal can coexist with both positive and negative correlation.

Exercise 3.95. Show that a mixture of two NWU distributions is NWU.

Exercise 3.96. Give a simple proof of Proposition 3.65 using the simplifying condition that $\overline{G}(t|\theta)$ is stochastically increasing in θ.

3.6 Bibliography

Arjas, E. (1981a). A stochastic process approach to multivariate reliability systems: notions based on conditional stochastic order. *Math. Op. Res.* 6., No. 2, 263–276.

Arjas, E. (1981b). The failure and hazard processes in multivariate reliability systems. *Math. Oper. Res.* 6, 551-562.

Arjas, E. (1989). Survival models and martingale dynamics. *Scand. J. Statist.* 16, 177-225.

Arjas, E. and Norros, I. (1984). Lifelengths and association: a dynamical approach. *Math. Op. Res.* 9, 151-158.

Arjas, E. and Norros, I. (1986). A compensator representation of multivariate life length distributions, with applications. *Scan. J. Stat.* 13, 99-112.

Arjas, E. and Norros, I. (1991). Stochastic order and martingale dynamics in multivariate life length models: a review. In *Stochastic Orders and Decision under Risk*, K. Mosler and M. Scarsini, Eds. Inst. Math. Statist., Hayward, CA.

Aven, T. and Jensen, U. (1999). *Stochastic Models in Reliability. Series Application of Mathematics.* Springer Verlag, New York.

Barlow, R. E. (1985). A Bayesian explanation of an apparent failure rate paradox. *IEEE Trans. on Rel.*, R34, No. 2, 107-108.

Barlow, R. E. and Proschan, F. (1966). *Mathematical Theory of Reliability.* Classics in Applied Mathematics, SIAM, New York.

Barlow, R. E. and Proschan, F. (1975). *Statistical Theory of Reliability and Life Testing.* Holt, Rinehart and Winston, New York.

Block, H.W. and Savits, T.H. (1997). Burn in. *Statistical Science*, 12, 1-19.

Block, H. W., Savits, T. H. and Shaked, M. (1982). Some concepts of negative dependence. *Ann. Probab.*, 10, 765-772.

Block, H. W., Sampson, A. and Savits, T.H. (Eds.) (1991). *Topics in Statistical Dependence.* Institute of Mathematical Statistics, Hayward, Ca.

Brindley, E.C. and Thompson, W.A. (1972) Dependence and aging aspects of multivariate survival. *JASA*, 67, 822-830.

Caramellino, L. and Spizzichino, F. (1996). WBF property and stochastic monotonicity of the Markov process associated to Schur-constant survival functions. *J. Multivariate Anal.*, 56, 153-163.

Çinlar, E., Shaked, M. and Shantikumar J.G. (1989). On lifetimes influenced by a common environment. *Stoch. Proc. Appl.* 33, 347-359.

Clarotti, C.A. and Spizzichino, F. (1990). Bayes burn-in decision procedures. *Probab. Engrg. Inform. Sci.*, 4, 437-445.

Cohen, J.E. (1986). An uncertainty principle in demography and the unisex issue. *American Statistician*, 40, 32-39.

Ebrahimi, N. and Ghosh, M. (1981). Multivariate negative dependence. *Comm. Statist.* A 10, 307-337.

Esary, J.D. and Proschan, F. (1970). A reliability bound for systems of maintained, interdependent components. *JASA*, 65, 329-338.

Esary, J.D., Marshall, A.W. and Proschan, F. (1973). Shock models and wear process. *Ann. Probab.* 1, 627-649.

Esary, J.D., Proschan, F. and Walkup, D.W. (1967). Association of random variables with applications. *Ann. Math. Stat.*, 38, 1466-1474.

Fahmy S., de B. Pereira C., Proschan F. and Shaked M. (1982). The influence of the sample on the posterior distribution. *Comm. Statist. A,* 11, 1757-1768.

Fortuin, C.M. , Ginibre, J. and Kasteleyn, P.W. (1971). Correlation inequalities on some partially ordered set. *Comm. Math. Phys.* 22, 89-103.

Gurland, J. and Sethuraman, J. (1994). Reversal of increasing failure rates when pooling failure data. *Technometrics*, 36, 416-418.

Gurland, J. and Sethuraman, J. (1995). How pooling failure data may reverse increasing failure rates. *J. Am. Statist. Assoc.*, 90, 1416-1423.

Harris, R. (1970). A multivariate definition for increasing failure rate distribution functions. *Ann. Math. Statist.*,37, 713-717.

Hougaard P. (1986). Survival models for heterogeneous populations derived from stable distributions. *Biometrika*, 73, 387-396.

Joag-Dev, K. and Kochar, S. (1996). A positive dependence paradox and aging distributions. *J. Indian Statist. Assoc.* 34, 105–112.

Joag-Dev, K. and Proschan, F. (1983). Negative association of random variables, with applications. *Ann. Statist.*, 11, 286-295.

Joag-Dev, K. and Proschan, F. (1995). A general composition theorem and its applications to certain partial orderings of distributions. *Stat. Prob. Lett.*, 22, 111-119.

Jogdeo K. (1978). On a probability bound of Marshall and Olkin. *Ann. Statist.*, 6, 232-234.

Karlin, S. (1968). *Total positivity*. Stanford University Press, Stanford, Ca.

Karlin, S. and Rinott, J. (1980a). Classes of orderings measures and related correlation inequalities. I multivariate totally positive distributions, *J. Mult. Analysis*, 10, 467-498.

Karlin, S. and Rinott, J. (1980b). Classes of orderings measures and related correlation inequalities. II multivariate reverse rule distributions. *J. Mult. Analysis*, 10, 499-516.

Karlin, S. and Rubin, H. (1956). The theory of decision procedures for distributions with monotone likelihood ratio. *Ann. Math. Statist.* 27, 272-299.

Keilson, J. and Sumita, U. (1982). Uniform stochastic orderings and related inequalities. *Canad. J. Statist.*, 10,181-189.

Kemperman, J.H.B. (1977). On the FKG-inequality for measures on a partially ordered space. *Indagationaes Mathematicae*, 13, 313-331.

Kimeldorf, G. and Sampson, A. R. (1989). A framework for positive dependence. *Ann. Inst. Math. Statist.*,41, 31-45.

Kochar, S. (1999). On stochastic orderings between distributions and their sample spacings. *Statist. Probab. Lett.*, 42 , 345-352.

Lehmann, E. L. (1966). Some concepts of dependence. *Ann. Math. Statist.* 37, 1137-1153.

Ligget, T. (1985). *Interacting Particle Systems.* Springer Verlag, New York.

Lynch, J. D. (1999). On conditions for mixtures of increasing failure rate distributions to have an increasing failure rate. *Probab. Engrg. Inform. Sci.*, 13, 33-36.

Lynch, J. D., Mimmack, G. and Proschan, F. (1987). Uniform stochastic orderings and total positivity. *Canad. J. Statist.*, 15, 63-69.

Manton, K.G., Stallard, E. and Vaupel, J.W. (1979). The impact of heterogeneity in individual frailty on the dynamics of mortality. *Demography*, 16, 439-454.

Marshall, A.V. (1975). Multivariate distributions with monotone hazard rate. In *Reliability and Fault Tree Analysis.* SIAM, Philadelphia, 259-284.

Mosler, K. and Scarsini, M. (Eds.) (1991). Stochastic orders and decision under risk. Institute of Mathematical Statistics. Hayward, Ca.

Mosler, K. and Scarsini, M. (1993) Stochastic Orders and Applications, A Classified Bibliography. Springer-Verlag, New York.

Nelsen, R. B. (1999). *An introduction to copulas.* Springer Verlag, New York.

Norros, I. (1985). Systems weakened by failures. *Stoch. Proc. Appl.*, 20, 181-196.

Norros, I. (1986). A compensator representation of multivariate life length distributions, with applications. *Scand. J. Statist.* 13, 99-112.

Pellerey, F. and Shaked, M. (1997). Characterizations of the IFR and DFR aging notions by means of the dispersive order. *Statist. Probab. Lett.* 33, 389–393.

Preston, C.J. (1974). A generalization of the KFG inequalities. *Comm. Math. Phys.*, 36, 223-241.

Proschan, F. (1963). Theoretical explanation of observed decreasing failure rate. *Technometrics*, 5, 375-383.

Ross, S. M. (1984). A model in which components failure rates depend only on the working set. *Nav. Res. Log. Quart.*, *31, 297-300.*

Samuels, M. L. (1993). Simpson's paradox and related phenomena. *J. Am. Statist. Assoc.*, 88 , no. 421, 81–88.

Sarkar, T.K. (1969). *Some lower bounds of reliability.* Tech. Report, No 124, Dept of Operations Research and Statistics, Stanford University.

Savits, T. H. (1985). A multivariate IFR distribution. *J. Appl. Probab.* 22, 197-204.

Scarsini, M. and Spizzichino, F. (1999). Sympson-type paradoxes, dependence and aging. *J. Appl. Probab.*, 36, 119-131

Shaked, M. (1977). A concept of positive dependence for exchangeable random variables. *Ann. Statist.*, 5, 505-515.

Shaked, M. and Scarsini, M. (1990). Stochastic ordering for permutation symmetric distributions. *Stat. Prob. Lett.*, 9, 217-222.

Shaked, M. and Shanthikumar, J.G. (1987a). Multivariate hazard rates and stochastic ordering. *Adv. Appl. Probab.* 19, 123-137.

Shaked, M. and Shanthikumar, J.G. (1987b). The multivariate hazard construction. *Stoc. Proc. Appl.*, 24, 85-97.

Shaked, M. and Shanthikumar, J.G. (1988). Multivariate conditional hazard rates and the MIFRA and MIFR properties. *J. Appl. Probab.*, 25, 150-168.

Shaked, M. and Shanthikumar, J. (1990). Multivariate stochastic orderings and positive dependence in reliability theory. *Math. Op. Res.* 15, 545-552.

Shaked, M. and Shanthikumar, J.G. (1991a). Dynamic multivariate aging notions in reliability theory. *Stoc. Proc. Appl.*, 38, 85-97.

Shaked, M. and Shanthikumar, J.G. (1991b). Dynamic construction and simulation of random vectors. In *Topics in Statistical Dependence*, H.W. Block, A. Sampson and T.H. Savits Eds., Institute of Mathematical Statistics, 415-433.

Shaked, M. and Shanthikumar, J.G.(1994). *Stochastic orders and their applications.* Academic Press, London.

Shaked, M. and Spizzichino, F. (1998). Positive dependence properties of conditionally independent random times. *Math. Oper. Res.*, 23, 944-959.

Shaked, M. and Spizzichino, F. (2001). Mixtures and monotonicity of failure rate functions. In *Handbook of Statistics: Advances in Reliability,* C. R. Rao and N. Balakrishnan, Eds. Elsevier, Amsterdam. To appear.

Simpson, E. H. (1951). The interpretation of interaction in contingency tables. *J. Roy. Statist. Soc.*, 13, 238-241.

Strassen, V. (1965). The existence of probability measures with given marginals. *Ann. Math. Statist.*, 36, 423-439.

Szekli, R. (1995). *Stochastic ordering and dependence in applied probability.* Lecture Notes in Mathematics 97, Springer Verlag, New York.

Tong, Y.L. (Ed.) (1991). *Inequalities in Statistics and Probability.* Institute of Mathematical Statistics. Hayward, Ca..

Vaupel, J.V. and Yashin, A.I. (1985). Some surprising effects of selection on population dynamics. *The Am. Statist.*, 39, 176-185.

Veinott, A.F. (1965). Optimal policy in a dynamic, single product, nonstationary inventory model with several demand classes. *Operat. Res.*, 13, 761-778.

Whitt, W. (1979). A note on the influence of the sample on the posterior distribution. *J. Am. Statist. Assoc.*, 74,424-426.

Whitt, W. (1980). Uniform conditional stochastic order. *J. Appl. Prob.* 17, 112-123.

Whitt, W. (1982). Multivariate monotone likelihood ratio and uniform conditional stochastic order. *J. Appl. Probab.* 19, 695-701.

Chapter 4

Bayesian models of aging

4.1 Introduction

In this chapter we introduce notions of multivariate aging, that we call "Bayesian", for reasons which will be clarified below. A discussion on motivations for introducing such notions will be presented here; we shall also analyze some structural differences with those reported in the previous chapter.

On this beginning, we however emphasize the basic aspects that Bayesian notions will have in common with the other notions:

a) the definitions are formulated by means of concepts of stochastic orderings

b) they concern conditional probability distributions of residual lifetimes of surviving individuals, given some survival and failure data for interdependent individuals

c) they arise in a natural way as extensions of inequalities respectively holding for independent lifetimes, under corresponding conditions of one-dimensional aging.

A specific feature of Bayesian notions, is, on the other hand, that they are essentially formulated for the cases of exchangeable lifetimes; we shall discuss briefly possibilities of extensions to other cases at the end of Section 4.4.

In order to explain the origin and the spirit of such notions, we start by noticing that various one-dimensional notions, such as IFR, DFR, and NBU, recalled in Section 3.4, can be formulated as conditions of the following type:

$$\mathcal{L}(T - t_1 | T > t_1) \geq_* \mathcal{L}(T - t_2 | T > t_2), \forall (t_1, t_2) \in A \qquad (4.1)$$

by suitably fixing $\geq_*$ as one of the one-dimensional stochastic ordering $\geq_{st}, \geq_{hr}$, $\geq_{lr}$ and fixing a subset $A \subset [0, +\infty) \times [0, +\infty)$.

For instance the concept of IFR can be obtained by letting $A \equiv \{(t_1, t_2) | t_1 \leq t_2\}$ and replacing $*$ with st or hr in (4.1).

As a second step, we now convert the relation (4.1) into an equivalent inequality for a set of independent lifetimes $T_1, ..., T_n$ distributed as T. In other words, we base our development on the following lemma, whose proof is trivial.

Lemma 4.1. *Let $T_1, ..., T_n$ be i.i.d. lifetimes, separately satisfying the inequality (4.1) and let D be an event of the form*

$$D \equiv \{(T_1 > t_1) \cap (T_2 > t_2) \cap H\} \tag{4.2}$$

with $(t_1, t_2) \in A$ and

$$H \equiv \{T_3 > t_3, ..., T_k > t_k, T_{k+1} = t_{k+1}, ..., T_n = t_n\}.$$

Then

$$\mathcal{L}(T_1 - t_1 | D) \geq_* \mathcal{L}(T_2 - t_2 | D) \tag{4.3}$$

The idea lies now in replacing the condition that $T_1, ..., T_n$ are i.i.d with the condition that the joint survival function $\overline{F}^{(n)}$ of $T_1, ..., T_n$ is exchangeable and taking then (4.3) as a notion of aging for $\overline{F}^{(n)}$.

This gives rise to several notions, according to the particular choice of A, and $\geq_*$, and, possibly, also according to the number of failures in the event H being positive or zero.

This line is developed in Bassan and Spizzichino (1999), where sufficient conditions for different inequalities of the type (4.3) are shown.

It originates from previous consideration of the following instance: as a particular multivariate notion of IFR for exchangeable lifetimes we can take the one that arises from imposing (4.3) with

$$A \equiv \{(t_1, t_2) | t_1 \leq t_2\}, \quad \geq_* = \geq_{st}, \quad \text{and} \quad k = n - 2.$$

In the next Section 4.2, we shall discuss in detail the fact that the latter yields a very natural notion, which is equivalent to the condition of *Schur-concavity* of the joint survival function; indeed the latter was already considered as a notion of aging (Barlow and Mendel, 1992).

In general the concept of Schur-concavity has an important role here, as will be explained in Section 4.2.

Section 4.3 will be specifically devoted to cases of absolute continuity; there, we analyze the meaning of Schur properties for the joint density function of exchangeable lifetimes.

In Section 4.4 we discuss some further aspects and details about notions of aging studied here.

As an introduction to the more technical treatment presented in the subsequent sections, we here point out some general aspects of definitions originating from (4.3) (and we thus trace some differences with Definitions 3.70-3.72).

First of all we notice that the inequality (4.3) regards the conditional distributions of two residual lifetimes given observed survival and failure data and that it is not of a dynamic type; in fact different survival data are contained in the observation, i.e., we compare the conditional distributions of two different individuals, of two different ages.

While in the inequalities (3.28)-(3.30), on the contrary, all surviving individuals share the same age, we may need, in some decision problems, to deal with individuals who start working (or living) in different time-instants (see in particular Remark 2.24).

The definitions 3.70-3.72 actually seem to be specifically appropriate for those cases where random variables are thought of as lifetimes of devices working simultaneously as different components of the same system.

This aspect is also related to a further, main, source of difference between the two types of definitions; the difference regards the kind of interdependence among individuals. In particular conditions of positive dependence, adapted to describe interactions among devices working simultaneously as different components of the same system, can be substantially different from dependence due to "conditional independence", as Remark 3.67 shows.

As a matter of fact, we want definitions of aging which are also compatible with correlations due to situations of learning about some unknown parameter. In this respect, we say that definitions of multivariate aging originating from (4.3) are "Bayesian". More generally we want definitions compatible with significant cases of exchangeability.

As remarked in Section 3.4, we must admit that the validity of a given aging property may depend on the actual flow of information. In fact the validity of some aging property can be destroyed in the cases when the effects of a difference in information overlap the effects of different ages for surviving individuals. This is just the case for Definitions 3.70-3.72, where a comparison is established between two conditional distributions having two different conditioning events.

However we want to face the following problem: *to single out inequalities which still remain unchanged under some special kinds of change of information, which do not destroy exchangeability.*

This is also explained by the following Remark.

Remark 4.2. As we saw in Chapter 1, the importance of the notion of exchangeability in Bayesian statistics is that it describes joint distributions of random quantities arising from *random sampling.*

In more probabilistic terms, we can say that exchangeability is a common property for the following different kinds of joint distribution

1) distributions of i.i.d. random variables,

2) distributions of conditionally i.i.d. random variables,

3) conditional distributions of originally i.i.d. random variables $T_1, ..., T_n$, given symmetric statistics such as $S_n \equiv \sum_{i=1}^{n} T_i$.

This means in particular that exchangeability is maintained under the operations of unconditioning and of conditioning with respect to symmetric statistics, whereas the condition of stochastic independence is in general lost.

In our specific setting, we more precisely start from lifetimes which are i.i.d. with monotone failure rates. Under the "Bayesian" operations mentioned above, even properties of monotonicity for one-dimensional failure rates can get lost along with stochastic independence.

In this respect we pursue multivariate aging notions which are maintained under such operations. The discussion in the following example may help clarify why we are interested in the problem formulated above.

Example 4.3. (The burn-in problem). Here we continue to comment on the burn-in problem discussed in Example 3.69, in the last chapter.

There, we assumed that, in deciding about possible burn-in of the unit U, there is no opportunity to observe further lifetimes affected by the same unknown parameter Θ and we realized that one-dimensional aging properties of conditional distributions $G(s|\theta)$ have no direct impact on the decision: only one-dimensional aging properties of the "predictive" survival function

$$\int_0^\infty \overline{G}(s|\theta)\pi(\theta)d\theta$$

are relevant to the problem.

Here we consider, on the contrary, the case when an individual I may use or test n units $U_1, ..., U_n$ with lifetimes $T_1, ..., T_n$ and he/she assesses that $T_1, ..., T_n$ are conditionally independent given Θ, with the same conditional survival function $\overline{G}(s|\theta)$.

In such a case I can choose among many more decision procedures than those envisaged in the single unit case; for instance I can decide to test a few units up to their failures in order to learn about Θ, so as to predict the behavior of the remaining units, without burning them in.

In the choice among different decision procedures (How many units to test? Which rule should be adopted to stop testing?), π and $\overline{G}(s|\theta)$ are to be taken into account.

Two individuals I_1, I_2, respectively, making probability assessment through π_i and $\overline{G}_i(s|\theta)$ $(i = 1, 2)$, may take different decisions even if the identity in (3.26) holds.

In particular one-dimensional aging properties of $\overline{G}_i(s|\theta)$ $(i = 1, 2)$ can have a direct impact on their decisions.

Then we may wonder:

"How the condition of $\overline{G}_i(s|\theta)$ being IFR (or DFR) for any θ reflects in the joint probability model for $T_1, ..., T_n$?"

or, better:

"Which inequalities are inherited by the joint (predictive) probability distributions of $T_1, ..., T_n$ in the case of monotonicity for the conditional hazard rate functions of $\overline{G}_i(s|\theta)$?"

Definitions of the type (4.3) actually aim to provide solutions to the above problem; an essential feature in fact is that they are only based on comparisons between conditional probabilities for different events under the same conditioning.

In this respect we shall see, for instance, that definitions of positive aging of the type (4.3) can encompass both the cases of i.i.d. lifetimes with IFR distributions and conditionally i.i.d. lifetimes with conditional IFR distributions, as well as other cases of exchangeable lifetimes originating from the condition of one-dimensional IFR (see in particular Remarks 4.19 and 4.41).

A further aspect to point out is the following: we shall see that the notions originating from (4.3) allow a certain compatibility between positive aging and negative dependence, as well as compatibility between negative aging and positive dependence.

Sometimes this may be appropriate, especially in the case of biological applications. For instance we can have in practice situations of negative dependence which are compatible with reasonable heuristic ideas of positive aging (see also the discussion in Brindley and Thompson, 1972).

Finally we note that in terms of (4.3), we can formulate multivariate notions of positive aging and of negative aging in a symmetric way.

From a technical point of view, we notice that an inequality of the type (4.3) is written in terms of one-dimensional stochastic orderings (while definitions 3.70-3.72 involve notions of multivariate stochastic orderings).

Some applications of the notions studied here to problems of Bayesian decisions will be indicated in the next chapter.

4.2   Schur survival functions

In this section we analyze the case of a vector of lifetimes $T_1, ..., T_n$ with an exchangeable survival function $\overline{F}^{(n)}$ such that the following condition holds:

$$P\{T_1 > s_1 + \tau | D\} \geq P\{T_2 > s_2 + \tau | D\}, \forall \tau > 0 \qquad (4.4)$$

where $s_1 < s_2$ and D is an event of the particular form

$$D \equiv \{(T_1 > s_1) \cap (T_2 > s_2) \cap (T_3 > s_3)... \cap (T_n > s_n)\}.$$

The condition (4.4) is equivalent to (4.3) with the particular choice $\geq_* = \geq_{st}$, $k = n - 2$.

As remarked in Section 4.1, the inequality (4.4) is equivalent to $T_1, ..., T_n$ being IFR in the case when $T_1, ..., T_n$ are i.i.d..

In the general case of exchangeability, it can still be seen as a condition of positive aging in the following sense: it asserts that, between two individuals which survived a life test, the "younger" is deemed to have higher probability than the "elder" to survive an extra time τ.

Notice that in (4.4), we do not compare the conditional distribution of a residual lifetime for an individual with the analogous conditional distribution for the residual lifetime of the same individual considered when "older" than now. Actually we compare conditional distributions of residual lifetimes of two individuals of different ages, given the same state of information.

We already noticed that this allows us to avoid undue effects of different states of information when comparing two conditional probabilities of survivals.

As mentioned, the condition (4.4) is strictly related to the concept of Schur-concavity. This will become clear after recalling the definitions of majorization, Schur-convexity, Schur-concavity and some related properties.

4.2.1   Basic background about majorization

In this subsection we recall the basic definitions of majorization ordering and of the related notion of Schur-convex and Schur-concave functions. A general reference for these topics is the book by Marshall and Olkin, 1979.

For a vector $\mathbf{x} \equiv (x_1, ..., x_n)$ of $\mathbb{R}^n$, denote by $x_{[1]}, ..., x_{[n]}$ the components of $\mathbf{x}$, rearranged in decreasing order: $x_{[1]} \geq ... \geq x_{[n]}$ and let $\mathbf{a} \equiv (a_1, ..., a_n)$ and $\mathbf{b} \equiv (b_1, ..., b_n)$ be two vectors of $\mathbb{R}^n$.

Definition 4.4. The vector $\mathbf{a}$ is greater than $\mathbf{b}$ in the majorization ordering (written $\mathbf{a} \succ \mathbf{b}$), if

$$\sum_{i=1}^{k} a_{[i]} \geq \sum_{i=1}^{k} b_{[i]}, k = 1, 2, ..., n-1 \text{ and } \sum_{i=1}^{n} a_i = \sum_{i=1}^{n} b_i.$$

Denoting by $x_{(1)}, ..., x_{(n)}$ the components of a vector $\mathbf{x}$, rearranged in an increasing order, the condition $\sum_{i=1}^{k} a_{[i]} \geq \sum_{i=1}^{k} b_{[i]}$ $(k = 1, 2, ..., n-1)$ can be rewritten as

$$\sum_{i=1}^{h} a_{(i)} \leq \sum_{i=1}^{h} b_{(i)}, \quad h = 1, 2, ..., n-1.$$

Example 4.5. In particular

$$(1, 0, ..., 0) \succ \left(\frac{1}{2}, \frac{1}{2}, 0, ..., \right) \succ$$

$$... \succ \left(\frac{1}{n-1}, ..., \frac{1}{n-1}, 0 \right) \succ \left(\frac{1}{n}, ..., \frac{1}{n} \right).$$

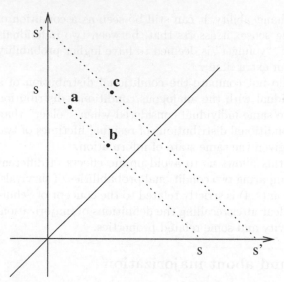

Figure 7. $\mathbf{a} \succ \mathbf{b}$; $\mathbf{c}$ is not comparable with $\mathbf{a}$ nor with $\mathbf{b}$, in the sense of majorization.

In general, starting from any vector $\mathbf{a} \equiv (a_1, ..., a_n)$, we can create a second vector $\mathbf{a}' \equiv (a_1', ..., a_n')$ such that $\mathbf{a} \succ \mathbf{a}'$; this can be done by means of the following procedure: all components of $\mathbf{a}'$ are equal to those of $\mathbf{a}$, but i-th coordinate and j-th coordinate, where a_i, a_j are respectively replaced by their convex combinations $(1 - \alpha) a_i + \alpha a_j$ and $(1 - \alpha) a_j + \alpha a_i$, with $0 \leq \alpha \leq 1$. Note that this also means

$$a_i' = a_i + \delta_\alpha, a_j' = a_j - \delta_\alpha$$

with $\delta_\alpha = \alpha(a_j - a_i)$.

For a given vector $\mathbf{a} \equiv (a_1, ..., a_n)$, it will be convenient to introduce the symbol $G_\alpha^{(i,j)}(\mathbf{a})$ to denote the vector

$$G_\alpha^{(i,j)}(\mathbf{a}) \equiv (a_1, ..., a_i + \delta_\alpha, ..., a_j - \delta_\alpha, ..., a_n) \qquad (4.5)$$

A fundamental fact about majorization is the following: $\mathbf{a} \succ \mathbf{a}'$ if and only if $\mathbf{a}'$ can be obtained from $\mathbf{a}$ by means of a finite number of successive transformations of the type in (4.5), with

$$a_i < a_j, 0 \leq \delta_\alpha \leq a_j - a_i$$

We point out that the transformation $G_\alpha^{(i,j)}$ is usually termed "T transform" (e.g. in the book by Marshall and Olkin, 1979). We preferred to use a different notation for practical reasons (in particular to avoid possible confusion with the notions of "TTT process" and "TTT Transform" used in the reliability literature).

Definition 4.6. A function $g : \mathcal{A} \subset \mathbb{R}^n \to \mathbb{R}$ is *Schur-convex on* $\mathcal{A}$ if it is non-decreasing with respect to the *majorization ordering*, i.e.

$$\mathbf{a}, \mathbf{b} \in \mathcal{A}, \mathbf{a} \succ \mathbf{b} \Rightarrow g(\mathbf{a}) \geq g(\mathbf{b})$$

If $\mathbf{a} \succ \mathbf{b} \Rightarrow g(\mathbf{a}) \leq g(\mathbf{b})$, then g is *Schur-concave on* $\mathcal{A}$.

Since we are essentially interested in Schur properties for functions of lifetimes, we shall understand that we deal with Schur functions defined on $\mathbb{R}_+^n$, if not explicitly mentioned otherwise.

We then simply say that g is *Schur-convex (Schur-concave)* when g is Schur-convex (Schur-concave) on $\mathcal{A} \equiv \mathbb{R}^n$. We say that g is a *Schur function* if it is Schur-concave or Schur-convex.

It is immediate that any Schur function g must necessarily be permutation invariant since, for any permutation $(\pi_1, ..., \pi_n)$ of $\{1, 2, ..., n\}$, trivially

$$(a_1, ..., a_n) \succ (a_{\pi_1}, ..., a_{\pi_n}) \text{ and } (a_{\pi_1}, ..., a_{\pi_n}) \succ (a_1, ..., a_n)$$

In particular we say that g is *Schur-constant* if it is, at a same time, Schur concave and Schur convex; g is Schur constant if and only if, for a suitable function $\phi : \mathbb{R} \to \mathbb{R}$, one has

$$g(\mathbf{a}) = \phi \left(\sum_{i=1}^{n} a_i \right).$$

Remark 4.7. A function g is Schur-concave if and only if, for each constant c, the level set $A_c \equiv \{\mathbf{a} | g(\mathbf{a}) \geq c\}$ is such that its indicator function is Schur-concave, i.e. such that

$$\mathbf{y} \in A_c, \mathbf{y} \succ \mathbf{x} \Rightarrow \mathbf{x} \in A_c \qquad (4.6)$$

Similarly g is Schur-convex if and only if, for each constant c, the level set A_c is such that its indicator function is Schur-convex (i.e. $\mathbf{x} \in A_c, \mathbf{y} \succ \mathbf{x} \Rightarrow \mathbf{y} \in A_c$).

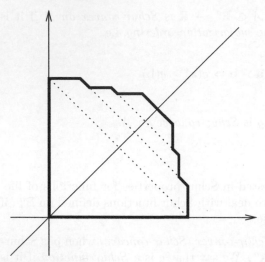

Figure 8. Example of a set with Schur-concave indicator function.

For our treatment we need some general properties of Schur functions, which will be reported next, directly adapting their formulation to the case $\mathcal{A} \equiv \mathbb{R}_+^n$.

Fix an ordered subset $I \subset \{1, 2, ..., n\}$ and denote, as usual, $\widetilde{I} \equiv \{1, 2, ..., n\} - I$.

For a given vector $\mathbf{t}_{\widetilde{I}} \in \mathbb{R}_+^{n-|I|}$, we now look at $g\left(\mathbf{t}_I, \mathbf{t}_{\widetilde{I}}\right)$ as a function of the vector $\mathbf{t}_I$.

Since it is immediate that $\mathbf{t}_I' \succ \mathbf{t}_I''$, implies $\left(\mathbf{t}_I', \mathbf{t}_{\widetilde{I}}\right) \succ \left(\mathbf{t}_I'', \mathbf{t}_{\widetilde{I}}\right)$, we obtain

Proposition 4.8. *Let $g : \mathbb{R}_+^n \to \mathbb{R}$ be a Schur-concave (Schur-convex) function then $g\left(\mathbf{t}_I, \mathbf{t}_{\widetilde{I}}\right)$ is Schur-concave (Schur-convex) as a function of $\mathbf{t}_I$, for any $I \subset \{1, 2, ..., n\}, \mathbf{t}_{\widetilde{I}} \in \mathbb{R}_+^{n-|I|}$.*

In the special case when $I \equiv \{i, j\}$, we shall write, for any given function g and for vectors
$$\widetilde{\mathbf{t}} \equiv (t_1, ..., t_{i-1}, t_{i+1}, ..., t_{j-1}, t_{j+1}, ..., t_n) \in \mathbb{R}_+^{n-2},$$

$$\gamma_g^{(i,j)}\left(t_i, t_j; \widetilde{\mathbf{t}}\right) \equiv g(t_1, ..., t_{i-1}, t_i, t_{i+1}, ..., t_{j-1}, t_j, t_{j+1}, ..., t_n) \qquad (4.7)$$

The following results provide useful characterizations of Schur functions (for proofs, see Marshall and Olkin, 1979, pg. 55-57).

Proposition 4.9. *A function g is Schur-concave (Schur-convex) if and only if g is permutation invariant and $\gamma_g^{(i,j)}(\xi, s - \xi; \mathbf{t})$ is increasing (decreasing) in $0 \le \xi \le \frac{s}{2}$ for each $s > 0$, and $\mathbf{t} \in \mathbb{R}_+^{n-2}$.*

The above conditions can be expressed in terms of derivatives when g is differentiable.

Proposition 4.10. *(Schur, 1923; Ostrowski, 1952). Let $g : (0, +\infty)^n \to \mathbb{R}$ be continuously differentiable. g is Schur-concave (or Schur-convex) if and only if it is permutation invariant and, for each i, j,*

$$(t_i - t_j) \cdot \left[\frac{\partial g}{\partial t_i}(t) - \frac{\partial g}{\partial t_j}(t) \right] \leq 0$$

$$\left(or \quad (t_i - t_j) \cdot \left[\frac{\partial g}{\partial t_i}(t) - \frac{\partial g}{\partial t_j}(t) \right] \geq 0 \right) \tag{4.8}$$

for all $t \in (0, +\infty)^n$.

The above inequalities are often called *Schur's conditions.*

The following properties are fundamental for our treatment.

Proposition 4.11. *If a symmetric function $g : \mathbb{R}^n \to \mathbb{R}$ is log-concave (log-convex) then it is Schur-concave (Schur-convex).*

Proposition 4.12. *If $g(\mathbf{a})$ is of the form $g(\mathbf{a}) = \varphi(a_1) \cdot \ldots \cdot \varphi(a_n)$ then it is Schur-concave (Schur-convex) if and only if φ is log-concave (log-convex).*

4.2.2   Schur properties of survival functions and multivariate aging

The starting point for the discussion to be developed here is the relation between arguments above and the concepts of IFR and DFR.

First we recall that a lifetime T has a IFR (DFR) distribution if and only if its survival function $\overline{G}(t) \equiv P\{T > t\}$ is log-concave (log-convex).

Proposition 4.12 then yields

Corollary 4.13. *The joint survival function $\overline{F}^{(n)}(s_1, ..., s_n)$ of i.i.d. lifetimes $T_1, ..., T_n$ is Schur-concave (Schur-convex) if and only if the one-dimensional distribution of $T_1, ..., T_n$ is IFR (DFR).*

In view of Corollary 4.13 one regards the properties of IFR and DFR (for one-dimensional distributions over $[0, \infty)$) as properties of a joint survival function for n i.i.d lifetimes.

Recalling that $T_1, ..., T_n$ being IFR, in the case when they are i.i.d., is equivalent to the inequality (4.4), Corollary 4.13 can also be reformulated as follows: for i.i.d. lifetimes $T_1, ..., T_n$, the following three conditions are equivalent

a) the inequalities (4.4) hold

b) $\overline{F}^{(n)}(s_1, ..., s_n) \equiv P\{T_1 > s_1, ..., T_n > s_n\}$ is Schur-concave on $\mathbb{R}^n_+$

c) $T_1, ..., T_n$ are IFR

As seen above, it is of interest to consider the condition a) even in the general case of exchangeable distributions for $T_1, ..., T_n$. On the other hand, Corollary 4.13 suggests the study of lifetimes with a Schur-concave joint survival function.

Remark 4.14. Since, as noticed above, a Schur function is necessarily permutation invariant, vectors of lifetimes with Schur joint survival functions are exchangeable.

The following result points out that the equivalence between the conditions a) and b) is true for any exchangeable survival function on $\mathbb{R}_+^n$.

However, these are not in general equivalent to c) anymore, if $T_1, ..., T_n$ are not independent (recall the discussion in Section 3.4 and see Remark 4.17 below).

Again let $D \equiv \{T_1 > s_1, ..., T_n > s_n\}$

Proposition 4.15. *Let $\overline{F}^{(n)}(s_1, ..., s_n)$ be an exchangeable joint survival function. Then the following two conditions are equivalent.*
a) $0 \leq s_1 < s_2 \Rightarrow$

$$\Rightarrow P\{T_1 > s_1 + \tau | T_1 > s_1, ..., T_n > s_n\} \geq$$

$$P\{T_2 > s_2 + \tau | T_1 > s_1, ..., T_n > s_n\}, \forall \tau > 0$$

b) $\overline{F}^{(n)}(s_1, ..., s_n) \equiv P\{T_1 > s_1, ..., T_n > s_n\}$ is Schur-concave.

Proof. One has

$$P\{T_i > s_i + \tau | D\} = \frac{\overline{F}^{(n)}(s_1, ..., s_{i-1}, s_i + \tau, s_{i+1}, ..., s_n)}{\overline{F}^{(n)}(s_1, ..., s_n)}$$

Then the implication a) means that, for $0 \leq r_i < r_j$, we have

$$\overline{F}^{(n)}(r_1, ..., r_{i-1}, r_i + u, r_{i+1}, ..., r_n) \geq \overline{F}^{(n)}(r_1, ..., r_{j-1}, r_j + u, r_{j+1}, ..., r_n).$$

Since, for $r_i < r_j$,

$$(r_1, ..., r_{i-1}, r_i + u, r_{i+1}, ..., r_n) \succ (r_1, ..., r_{j-1}, r_j + u, r_{j+1}, ..., r_n),$$

it is immediate that, if $\overline{F}^{(n)}$ is Schur concave, then the above implication holds.

Vice versa, if the implication a) holds, then $\overline{F}^{(n)}$ is Schur-concave as a consequence of the characterization given in Proposition 4.9.   □

Remark 4.16. Proposition 4.15 provides motivations to interpret Schur-concavity of the joint survival function as a condition of multivariate aging.

By reversing inequalities, we can also immediately see that $\overline{F}^{(n)}$ is Schur-convex on $\mathbb{R}_+^n$ if and only if $T_1, ..., T_n$ are exchangeable and the implication

$$s_1 < s_2 \Rightarrow P\{T_1 > s_1 + \tau | D\} \leq P\{T_2 > s_2 + \tau | D\}, \forall \tau > 0$$

holds true. By arguments similar to those above, the latter implication can be interpreted as a condition of infant mortality.

A further aspect of positive aging exhibited by vectors of lifetimes with Schur-concave survival function can be obtained by noting the following: an immediate consequence of this assumption is

$$P\{T_1 > s_1, ..., T_n > s_n\} \geq P\{T_1 > t\}, \forall t \geq \sum_{i=1}^{n} s_i$$

In particular, by letting $s_1 = ... = s_n = u$, we obtain

$$P\left\{\min_{1 \leq i \leq n} T_i > u\right\} \geq P\left\{\frac{T_1}{n} > u\right\}$$

i.e.

$$\min_{1 \leq i \leq n} T_i \geq_{st} \frac{T_1}{n}.$$

The above inequalities are reversed in the case of Schur-convexity.

Remark 4.17. As already noticed, the property that one-dimensional marginal survival functions are IFR (i.e. log-concave) is in general lost under mixing (see Remark 3.66 in Section 3.4).

Consider in fact the joint survival function of lifetimes $T_1, ..., T_n$ which are conditionally i.i.d. given a random variable Θ:

$$\overline{F}^{(n)}(s_1, ..., s_n) = \int_L \overline{G}(s_1|\theta) \cdot ... \cdot \overline{G}(s_n|\theta) \, d\Pi_0(\theta),$$

with Π_0 a given probability distribution over L.

$\overline{G}(\cdot|\theta)$ being log-concave, $\forall \theta \in L$, (i.e. $T_1, ..., T_n$ IFR given $\{\Theta = \theta\}$) does not imply that the one-dimensional predictive survival function $\overline{F}^{(1)}(s) = \int_L \overline{G}(s|\theta) d\Pi_0(\theta)$ is log-concave as well. This shows that the one-dimensional marginal of a n-dimensional Schur-concave survival function is not necessarily IFR.

Here we remark that, on the contrary, the implication in a) remains valid under mixtures.

Indeed, consider a vector of n lifetimes $\mathbf{T} \equiv (T_1, ..., T_n)$, with conditional distributions, given the parameter Θ, such that $\forall \tau > 0$, $\forall \theta \in L$, and again with $D \equiv \{T_1 > s_1, ..., T_n > s_n\}$, one has

$$0 < s_i < s_j \Rightarrow$$

$$P\{T_i > s_i + \tau | D; \theta\} \geq P\{T_j > s_j + \tau | D; \theta\}.$$

Then

$$P\{T_i > s_i + \tau | D\} = \int_L P\{T_i > s_i + \tau | D; \theta\} d\Pi_0(\theta|D) \geq$$

$$\int_L P\{T_j > s_j + \tau | D; \theta\} d\Pi_0(\theta|D) = P\{T_j > s_j + \tau | D\}.$$

This same fact can be reformulated by means of the following statement

Proposition 4.18. *Let $\overline{F}^{(n)}$ be a joint survival function of the form*

$$\overline{F}^{(n)}(s_1, ..., s_n) = \int_L \overline{G}(s_1, ..., s_n | \theta) d\Pi(\theta)$$

with $\overline{G}(s_1,...,s_n | \theta)$ Schur-concave (or Schur-convex) $\forall \theta \in L$. Then $\overline{F}^{(n)}(s_1, ..., s_n)$ is Schur-concave (or Schur-convex).

Arguments presented above can be summarized as follows.

Remark 4.19. Schur-concavity (or Schur-convexity) of the joint survival function $\overline{F}^{(n)}$ is a property shared both by i.i.d. lifetimes with IFR (or DFR) distributions and by lifetimes which are conditionally i.i.d. IFR (or DFR).

The following results show further useful closure properties.

Proposition 4.20. *Let $\overline{F}^{(n)}(\mathbf{s})$ be a Schur-concave (Schur-convex) survival function. Then its k-dimensional marginal $\overline{F}^{(k)}$ is Schur-concave (Schur-convex) as well $(k = 2, ..., n-1)$.*

Proof. We note that

$$\overline{F}^{(k)}(s_1, ..., s_k) = \overline{F}^{(n)}(s_1, ..., s_k, 0, ..., 0)$$

Then the proof is immediately achieved by taking into account the implication

$$(s'_1, ..., s'_k) \succ (s''_1, ..., s''_k) \Rightarrow$$

$$(s'_1, ..., s'_k, 0, ..., 0) \succ (s''_1, ..., s''_k, 0, ...0).$$

$\square$

Fix an increasing transformation $\varphi : \mathbb{R}_+ \to \mathbb{R}_+$ and consider the transformed lifetimes $U_1, ..., U_n$ defined by

$$U_i \equiv \varphi(T_i), i = 1, ..., n$$

The joint survival function of $\mathbf{U} \equiv (U_1, ..., U_n)$ is given by

$$\overline{H}^{(n)}(u_1, ..., u_n) = P\{T_1 > \varphi^{-1}(u_1), ..., T_n > \varphi^{-1}(u_n)\}$$
$$= \overline{F}^{(n)}(\varphi^{-1}(u_1), ..., \varphi^{-1}(u_n)).$$

Proposition 4.21. *Let $\overline{F}^{(n)}$ be a Schur-convex joint survival function and let the function $\varphi : \mathbb{R}_+ \to \mathbb{R}_+$ be increasing and convex. Then*

$$\overline{H}^{(n)}(u_1, ..., u_n) = \overline{F}^{(n)}(\varphi^{-1}(u_1), ..., \varphi^{-1}(u_n))$$

is also Schur-convex.

An analogous property holds for Schur-concave survival functions.

We direct the reader to the book by Marshall and Olkin (1979) for the proof of Proposition 4.21, details on compositions of Schur functions, and for many other closure properties which can be converted into closure properties of Schur distributions for lifetimes.

4.2.3   Examples of Schur survival functions

Example 4.22. (Proportional hazard models). Consider the model in Example 2.18, Example 2.40, Chapter 2, and Example 3.46, Chapter 3.

For any distribution Π_0 over $(0, +\infty)$ the corresponding joint survival function $\overline{F}^{(n)}$ is Schur-concave when the function $R(t)$ is convex, and it is Schur-convex when the function $R(t)$ is concave (see also Barlow and Mendel, 1992 and Hayakawa, 1993).

Example 4.23. (Proportional hazard models; Schur-constant case). In the proportional hazard model with $R(t) = t$, we obtain that joint survival function is Schur-constant. When the distribution Π_0 is degenerate on a value $\theta > 0$, we in particular obtain the case of i.i.d. exponential lifetimes:

$$\overline{F}^{(n)}(s_1, ..., s_n) = \exp\{-\theta \cdot \sum_{i=1}^{n} t_i\},$$

$$f^{(n)}(t_1, ..., t_n) = \theta^n \cdot \exp\{-\theta \cdot \sum_{i=1}^{n} t_i\}.$$

Example 4.24. (Time-transformed exponential models). Continuing Example 2.17, consider more generally, survival functions of the form

$$\overline{F}^{(n)}(s_1, ..., s_n) = H^{(n)}(R(s_1), ..., R(s_n)) = \overline{\Phi}\left(\sum_{i=1}^{n} R(s_i)\right) \qquad (4.9)$$

where R is a given increasing function and $H^{(n)}$ is a Schur-constant survival function.

Then $H^{(n)}$ is at a time Schur-concave and Schur-convex and $\overline{F}^{(n)}(s_1, ..., s_n)$ in (4.9) is Schur-concave, Schur-constant, or Schur-convex according to whether R is concave, linear, or convex.

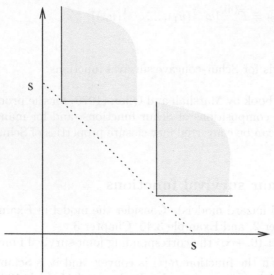

Figure 9. A Schur-constant survival function $\overline{F}^{(2)}$ is constant over the simplex Σ_s.

Example 4.25. From the definition of majorization ordering, it is immediate that $\mathbf{a} \succ \mathbf{b}$ implies $\max\{a_1, ..., a_n\} \geq \max\{b_1, ..., b_n\}$. Then the function

$$g(\mathbf{a}) \equiv \max\{a_1, ..., a_n\}$$

is Schur-convex, whence, for any non-increasing function $\phi : \mathbb{R} \to \mathbb{R}$, $\phi[\max\{a_1, ..., a_n\}]$ is Schur-concave.

In particular any joint survival function $\overline{F}^{(n)}(s_1, ..., s_n)$, which depend on $(s_1, ..., s_n)$ only through $\max\{s_1, ..., s_n\}$, must be Schur-concave.

This can be seen as an extreme case of Schur-concavity.

Consider the extreme cases of positive dependence where

$$P\{T_1 = \dots = T_n\} = 1.$$

Letting $\overline{F}^{(1)}$ denote the one-dimensional survival function of T_j ($j = 1, \dots, n$), the corresponding survival functions are of the form

$$\overline{F}^{(n)}(s_1, \dots, s_n) = \overline{F}^{(1)}(\max\{s_1, \dots, s_n\})$$

and are Schur-concave.

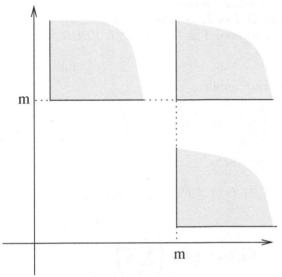

Figure 10 A level set of a survival function of the form $\overline{F}^{(2)}(s_1, s_2) = \phi(\max(s_1, s_2))$.

Example 4.26. (Common mode failures, bivariate model of Marshall-Olkin). The two-dimensional joint survival function

$$\overline{F}^{(2)}(s_1, s_2) = \exp\left\{-\lambda(s_1 + s_2) + \lambda' \max(s_1, s_2)\right\}$$

arising as a special case of the common-mode failure model considered in Example 2.23, is Schur-concave (see also Marshall and Olkin, 1974).

This property fits with the following heuristic explanation: suppose that one component is inspected at time s_1 and the other is inspected at the time s_2 and suppose that both are found to be alive. Then we know that there has been no shock before s_2 and hence our hopes of finding the first component still alive shortly after time s_1 are greater than hopes of finding the second component alive shortly after time s_2.

4.2.4 Schur survival functions and dependence

The condition of Schur-concavity for a survival function $\overline{F}^{(n)}$ can coexist with situations of independence, or positive dependence or negative dependence as well. The same happens for survival functions which are Schur-constant or Schur-convex (see in particular the above Example 4.23).

However, certain compatibility conditions among Schur properties and dependence properties can be obtained in terms of aging properties of one dimensional marginal distributions. For this purpose, recall the definitions of *NBU*, *NWU*, *PUOD* and *NUOD*, given in Chapter 3.

Proposition 4.27. *Let $\overline{F}^{(n)}$ be Schur concave.*

a) If $\mathbf{T} \equiv (T_1, ..., T_n)$ is NUOD then T_i is NBU $(i = 1, 2, ..., n)$.

b) If T_i is NWU $(i = 1, 2, ..., n)$ then $\mathbf{T} \equiv (T_1, ..., T_n)$ is PUOD

Proof. a) $\overline{F}^{(n)}$ being Schur concave implies

$$\overline{F}^{(n)}(s_1, ..., s_n) \geq \overline{F}^{(n)}\left(\sum_{i=1}^{n} s_i, 0, ..., 0\right) = \overline{F}_{T_i}\left(\sum_{i=1}^{n} s_i\right).$$

On the other hand, if $\mathbf{T}$ is NUOD then

$$\overline{F}_{T_i}(s_1) \cdot ... \cdot \overline{F}_{T_i}(s_n) \geq \overline{F}^{(n)}(s_1, ..., s_n)$$

whence

$$\overline{F}_{T_i}(s_1) \cdot ... \cdot \overline{F}_{T_i}(s_n) \geq \overline{F}_{T_i}\left(\sum_{i=1}^{n} s_i\right),$$

which in particular implies the NBU property.

b) T_i NWU means, for any $s, t \geq 0$,

$$\overline{F}_{T_i}(s + t) \geq \overline{F}_{T_i}(s)\overline{F}_{T_i}(t)$$

whence, if $\overline{F}^{(n)}$ is Schur concave

$$\overline{F}^{(n)}(s_1, ..., s_n) \geq \overline{F}_{T_i}\left(\sum_{i=1}^{n} s_i\right) \geq \overline{F}_{T_i}(s_1) \cdot ... \cdot \overline{F}_{T_i}(s_n).$$

$\square$

In a similar way one can prove

Proposition 4.28. *Let $\overline{F}^{(n)}$ be Schur convex.*

a') If $\mathbf{T} \equiv (T_1, ..., T_n)$ is PUOD then T_i is NWU $(i = 1, 2, ..., n)$.

b') If T_i is NBU $(i = 1, 2, ..., n)$ then $\mathbf{T} \equiv (T_1, ..., T_n)$ is NUOD.

Very special cases where $\overline{F}^{(n)}$ is Schur-convex and $\mathbf{T}$ is PUOD are those of lifetimes which are conditionally independent identically distributed, with a DFR (conditional) distribution, given some parameter. Then part a') of Proposition 4.28 can be seen as a generalization of the following well-known result, which gives a property for T_i that is stronger than NWU, under stronger dependence conditions among $T_1, ..., T_n$.

Proposition 4.29. *(Barlow and Proschan, 1975). If $T_1, ..., T_n$ are conditionally independent identically distributed, with a DFR (conditional) distribution, then T_i is DFR $(i = 1, 2, ..., n)$.*

4.3   Schur density functions

In the case of independence (and identical distribution), the inequality (4.4), which is equivalent to Schur-concavity of survival functions, is also equivalent to one in which the conditioning event can contain some *failure data* along with *survival data*, i.e., for i.i.d. IFR lifetimes, we obviously have, $\forall \tau > 0$, the validity of the following implication

$$s_1 < s_2 \Rightarrow$$

$$P\{T_1 > s_1 + \tau | T_1 > s_1, T_2 > s_2, ..., T_k > s_k, T_{k+1} = t_{k+1}, ..., T_n = t_n\} \geq \tag{4.10}$$

$$P\{T_2 > s_2 + \tau | T_1 > s_1, T_2 > s_2, , ..., T_k > s_k, T_{k+1} = t_{k+1}, ..., T_n = t_n\}$$

Such an equivalence is not true in general when stochastic dependence is present. We then turn now to conditions on an exchangeable survival function $\overline{F}^{(n)}$ which guarantee the validity of (4.10).

In order to avoid technical problems in considering conditioning events of the form

$$D \equiv \{T_1 > s_1, T_2 > s_2, , ..., T_k > s_k, T_{k+1} = t_{k+1}, ..., T_n = t_n\} \tag{4.11}$$

we assume that $\overline{F}^{(n)}$ admits a joint density $f^{(n)}$.

Below, we show that the validity of (4.10) is implied by Schur-concavity of $f^{(n)}$. Actually this is a stronger condition than Schur-concavity of the joint survival function, as can be seen in view of the following property.

For a set $A \subset \mathbb{R}_+^n$ and a vector $\mathbf{s} \in \mathbb{R}_+^n$, denote by $A + \mathbf{s}$, the set defined by

$$A + \mathbf{s} \equiv \{\mathbf{t} \in \mathbb{R}_+^n | \mathbf{t} = \mathbf{a} + \mathbf{s}, \text{ for some } \mathbf{a} \in A\}$$

so that the event $\{\mathbf{T} \in A + \mathbf{s}\}$ means $\{\mathbf{T} - \mathbf{s} \in A\}$.

Let now $A \subset \mathbb{R}_+^n$ have, in particular, a Schur-concave indicator function (i.e. an indicator function satisfying the implication (4.6)).

Theorem 4.30. *(Marshall and Olkin, 1974) If $f^{(n)}$ is Schur-concave on $\mathbb{R}^n_+$,* *then*

$$P\{\mathbf{T} \in A + \mathbf{s}\} = \int_{A+\mathbf{s}} f^{(n)}(t_1, ..., t_n) dt_1 ... dt_n$$

is a Schur-concave function of $\mathbf{s} \in \mathbb{R}^n_+$.

Note that, letting in particular $A \equiv \mathbb{R}^n_+$, it is

$$P\{\mathbf{T} \in A + \mathbf{s}\} = P\{T_1 > s_1, ..., T_n > s_n\} = \overline{F}^{(n)}(\mathbf{s}).$$

From Theorem 4.30 we can then obtain

Corollary 4.31. *If $f^{(n)}$ is Schur-concave on $\mathbb{R}^n_+$, then also the joint survival* *function $\overline{F}^{(n)}$ is Schur-concave on $\mathbb{R}^n_+$.*

If $f^{(n)}$ is Schur-convex on $\mathbb{R}^n_+$, then also the joint survival function $\overline{F}^{(n)}$ is *Schur-convex on $\mathbb{R}^n_+$.*

In the following we shall generally formulate results only for the case of Schur-concavity.

We recall that, by Proposition 4.11, the condition that $f^{(n)}$ is Schur-concave is in particular verified when $f^{(n)}$ is symmetric and log-concave.

Lemma 4.32. *Let $f^{(n)}$ be a Schur-concave joint density for a vector of life-* *times $T_1, ..., T_n$. The conditional survival function of $T_1, ..., T_k$ given*

$$H \equiv \{T_{k+1} = t_{k+1}, ..., T_n = t_n\}$$

is Schur-concave.

Proof. Taking into account the identity

$$f^{(k)}(\xi_1, ..., \xi_k | H) = \frac{f^{(n)}(\xi_1, ..., \xi_k, t_{k+1}, ..., t_n)}{f^{(n-k)}(t_{k+1}, ..., t_n)},$$

the proof is immediate, in view of the implication

$$f^{(n)} \text{ Schur-concave} \Rightarrow$$

$f^{(n)}(\xi_1, ..., \xi_k, t_{k+1}, ..., t_n)$ Schur-concave as a function of $(\xi_1, ..., \xi_k)$.

□

By simply applying this Lemma, Corollary 4.31, and using Proposition 4.15 for the variables $T_1, ..., T_k$, we obtain

Proposition 4.33. *Let $f^{(n)}$ be a Schur-concave joint density for a vector of lifetimes $T_1, ..., T_n$. Then, $\forall \tau > 0$, the implication (4.10) holds.*

Similarly to the above, one can also easily obtain, for an observation D as in (4.11),

Remark 4.34. If $f^{(n)}$ is Schur-convex, then the following implication holds $\forall \tau > 0$:

$$s_1 < s_2 \Rightarrow P\{T_1 > s_1 + \tau | D\} \leq P\{T_2 > s_2 + \tau | D\}. \tag{4.12}$$

Later in this section we shall consider different aspects of vectors of lifetimes admitting a joint density $f^{(n)}(t_1, ..., t_n)$ which is a Schur function.

Before, we present a few examples of Schur densities; we start by discussing in some detail the special case of Schur-constant density functions.

4.3.1   Schur-constant densities

Consider the case of lifetimes $T_1, ..., T_n$ with a Schur-constant joint density $f^{(n)}$:

$$f^{(n)}(t_1, ..., t_n) = \phi_n \left(\sum_{i=1}^{n} t_i \right). \tag{4.13}$$

We already noticed that $f^{(n)}$ being Schur-constant implies that $\overline{F}^{(n)}$ is Schur-constant as well. This can also be immediately obtained by taking into account that, in such a case, $f^{(n)}$ is simultaneously Schur-concave and Schur-convex and then applying the above Corollary 4.31.

By differentiation, one can easily check, vice versa, that if $\overline{F}^{(n)}$ is absolutely continuous and Schur-constant then also the density is Schur-constant; then an absolutely continuous $\overline{F}^{(n)}$ is Schur-constant if and only if its density $f^{(n)}$ is such.

More precisely we have that the condition (4.13) is equivalent to

$$\overline{F}^{(n)}(s_1, ..., s_n) = \overline{\Phi} \left(\sum_{i=1}^{n} s_i \right)$$

with $\overline{\Phi} : \mathbb{R}_+ \rightarrow [0, 1]$ such that $\phi_n(t) = (-1)^n \frac{d^n}{dt^n} \overline{\Phi}(t)$.

Notice that the function $\overline{\Phi}$ is a non-increasing function with the meaning of one-dimensional survival function of a single lifetime T_i $(i = 1, 2, ...,)$; in fact

$$\overline{F}^{(1)}(s) = \overline{F}^{(n)}(s, 0, ..., 0) = \overline{\Phi}(s).$$

Remark 4.35. In Example 2.35 of Chapter 2 we saw that the condition (4.13) also implies that the corresponding m.c.h.r. functions $\lambda_t^{(n-h)}(t_1, ..., t_h)$ ($h = 2, ..., n-1$) are Schur-constant functions of $t_1, ..., t_h$.

The condition (4.13) is trivially verified in the cases of independent, identically distributed, exponential lifetimes, where

$$\phi_n(t) = \theta^n \exp\{-\theta t\} \tag{4.14}$$

and of conditionally independent, identically distributed, exponential lifetimes.

Several probabilistic properties which hold for the case (4.14) also hold for arbitrary Schur-constant densities.

An example of that is provided by the significant invariance property, concerning conditional distributions of residual lifetimes, that we are going to illustrate now.

From Proposition 4.33 and Proposition 4.34, we obtain that, under the condition (4.13), the implications (4.10) and (4.12) simultaneously hold, and then we have

$$P\{T_1 > s_1 + \tau | D\} = P\{T_2 > s_2 + \tau | D\} \tag{4.15}$$

for any pair $s_1, s_2 > 0$ and for D in (4.11).

For the case of i.i.d lifetimes, Equation (4.15) is nothing but the "memoryless" property of exponential distributions.

Remark 4.36. More generally, in view of Equation (4.15), the condition (4.13) can be seen as the proper "multivariate" formulation of indifference with respect to age: given the same observation D, the conditional probability of extra survival is the same for any surviving individual, irrespective of its own age.

Such a property of Schur-constant densities, which could be called "no-aging", was also pointed out by Barlow and Mendel (1992).

We notice that something (apparently) stronger happens under the condition (4.13). Consider the conditional density

$$f_{T_1-s_1,T_2-s_2,...,T_k-s_k}(x_1, ..., x_k | D)$$

of residual lifetimes $T_1 - s_1, T_2 - s_2, ..., T_k - s_k$, conditional on the observation D.

Under the condition that a joint density does exist, we actually have

Proposition 4.37. *The following conditions are equivalent:*

(a) The joint density $f^{(n)}$ is Schur-constant

(b) $f_{T_1-s_1,T_2-s_2,...,T_k-s_k}(x_1, ..., x_k | D)$ is Schur-constant, $\forall s_1, ..., s_k, t_{k+1}, ..., t_n$.

(c) $f_{T_1-s_1,T_2-s_2,...,T_k-s_k}(x_1, ..., x_k | D)$ is exchangeable, $\forall s_1, ..., s_k, t_{k+1}, ..., t_n$.

(d) Given D, $T_1 - s_1, T_2 - s_2, ..., T_k - s_k$ are identically distributed, $\forall s_1, ..., s_k$, and $\forall t_{k+1}, ..., t_n$.

Proof. By adapting Proposition 2.15 to the present notation and applying it to the case when (a) holds, one can easily check that (b) holds.

Obviously (b) implies (c) which in turn implies (d).

In view of Proposition 4.15 and Remark 4.16, (d) implies that the joint survival function must be Schur-constant. As noticed, when a joint density does exist, the latter condition implies (a).   □

In conclusion (4.13) is equivalent to a condition of exchangeability (and then of indifference) for the residual lifetimes, which can be interpreted as a condition of indifference with respect to aging. We recall that combining such a condition with infinite extendibility we get, via Theorem 1.44, the "lack of memory" models defined by exponentiality or conditional exponentiality of lifetimes (go back to Remark 1.46).

Some other characterizations of Schur-constant densities follow from arguments to be developed next (see e.g. Remark 4.51). Further characterizations, which point out different aspects of such distributions, have been discussed by Chick and Mendel (1998).

Examples of lifetimes with Schur-constant densities which are not necessarily independent, or conditionally independent, and exponentially distributed (i.e. not necessarily infinitely extendible) can be produced as shown by the next example.

Example 4.38. Consider an arbitrary Schur-constant density

$$f^{(n)}(t_1, ..., t_n) = \varphi_n \left(\sum_{i=1}^{n} t_i \right)$$

and random variables $Y_1, ..., Y_m$ such that

$$(T_1, ..., T_n) \text{ and } (Y_1, ..., Y_m)$$

are conditionally independent, given $S_n \equiv \sum_{i=1}^{n} T_i$. This means that the conditional density

$$f_{\mathbf{Y}}(\mathbf{y}|T_1 = t_1, ..., T_n = t_n)$$

is a Schur-constant function of $t_1, ..., t_n$.

In this case, the conditional density

$$f_{\mathbf{T}}(\mathbf{t}|Y_1 = y_1, ..., Y_m = y_m) \propto$$

$$\propto \varphi_n \left(\sum_{i=1}^{n} t_i \right) f_{\mathbf{Y}}(\mathbf{y}|T_1 = t_1, ..., T_n = t_n)$$

is Schur-constant. Note that in general this is not a case of proportional hazard.

In Example 3.45, Chapter 3, we saw that we have the MTP$_2$ property if and only if ϕ_n is log-convex and this implies that the one-dimensional survival function $\overline{\Phi}$ is DFR; if $\overline{\Phi}$ is IFR we have a case of negative dependence.

This argument can be used to obtain, by means of the construction considered in Example 4.38, Schur-constant models which are not proportional hazard models (see Exercise 4.75).

In order to discuss further aspects of Schur densities, we shall need to come back to the case of Schur-constant densities, later on.

In particular, in Section 5.3, it will be necessary to point out further aspects of invariance, different from that shown by Proposition 4.37.

4.3.2   Examples of Schur densities

An elementary property of Schur-concavity and Schur-convexity is the one described by Proposition 4.18. It is obvious that an analogous implication is true for joint density functions; that is we have

Proposition 4.39. *Let $f^{(n)}$ be a joint survival function of the form*

$$f^{(n)}(t_1, ..., t_n) = \int_L g(t_1, ..., t_n|\theta)d\Pi_0(\theta)$$

with $g(t_1, ..., t_n|\theta)$ Schur-concave (or Schur-convex) $\forall \theta \in L$. Then $f^{(n)}(t_1, ..., t_n)$ is Schur-concave (or Schur-convex).

A different but, in a sense, similar closure property is described next. Let $f^{(n)}$ be a joint density for a vector of lifetimes $T_1, ..., T_n$ and denote by

$$f^{(n)}(t_1, ..., t_{n-1}|S_n = s)$$

the conditional density of $T_1, ..., T_{n-1}$ given $S_n = s$, where $S_n \equiv \sum_{i=1}^n T_i$.

Proposition 4.40. *If $f^{(n)}$ is Schur-concave (Schur-convex) then*

$$f^{(n-1)}(t_1, ..., t_{n-1}|S_n = s)$$

is Schur-concave (Schur-convex) as well.

Proof. We limit ourselves to considering the case of Schur-concavity. For $s > 0$, $(t_1, ..., t_{n-1}) \in \Delta_s^{(n-1)}$ where

$$\Delta_s^{(n-1)} \equiv \left\{ (t_1, ..., t_{n-1}) \in \mathbb{R}_+^{n-1} \mid \sum_{i=1}^{n-1} t_i \leq s \right\},$$

$$f^{(n-1)}(t_1, ..., t_{n-1}|S_n = s) = \frac{f^{(n)}(t_1, ..., t_{n-1}, s - \sum_{i=1}^{n-1} t_i)}{f_{S_n}(s)}$$

with

$$f_{S_n}(s) = \int_{\Delta_s^{(n-1)}} f^{(n)}(t_1, ..., t_{n-1}, s - \sum_{i=1}^{n-1} t_i) dt_1 ... dt_{n-1}.$$

Consider now two vectors $(t'_1, ..., t'_{n-1}), (t_1, ..., t_{n-1}) \in \Delta_s^{(n-1)}$ such that

$$\sum_{i=1}^{n-1} t'_i = \sum_{i=1}^{n-1} t_i = z, (t'_1, ..., t'_{n-1}) \succ (t_1, ..., t_{n-1})$$

If $f^{(n)}$ is Schur-concave, one then has, by Proposition 4.8,

$$f^{(n-1)}(t_1, ..., t_{n-1}|S_n = s) \geq \frac{f^{(n)}(t'_1, ..., t'_{n-1}, s)}{f_{S_n}(s)} =$$

$$f^{(n-1)}(t'_1, ..., t'_{n-1}|S_n = s).$$

□

Remark 4.41. Coming back to Remark 4.2, we summarize the arguments above as follows. Schur properties of joint densities are not lost under unconditioning nor under conditioning with respect to the statistic $S_n \equiv \sum_{i=1}^n T_i$.

Propositions 4.39 and 4.40 can be used to build several examples of Schur densities. In particular, recalling the Proposition 4.12, we get

Proposition 4.42. *Let Θ be a parameter taking values in a set L, with a prior distribution Π_0. Let $T_1, ..., T_n$ be conditionally i.i.d., given $\{\Theta = \theta\}$, with a conditional density $g(t|\theta)$:*

$$f^{(n)}(t_1, ..., t_n) = \int_L g(t_1|\theta)...g(t_n|\theta) d\Pi_0(\theta)$$

If $g(t|\theta)$ is log-concave (log-convex) then $f^{(n)}(t_1, ..., t_n)$ is Schur-concave (Schur-convex).

Example 4.43. (Conditionally i.i.d. Weibull lifetimes). Let Θ be a non-negative random variable with a prior distribution Π_0.

Let $T_1, ..., T_n$ be conditionally i.i.d., given $\{\Theta = \theta\}$ ($\theta > 0$), with Weibull density

$$g(t|\theta) = \theta \beta t^{\beta-1} \exp\{-\theta t^\beta\}, t \geq 0$$

where β is a positive constant.

$g(t|\theta)$ is log-concave if $\beta > 1$ and it is log-convex if $0 < \beta < 1$.

The joint density of $T_1, ..., T_n$ is then Schur-concave if $\beta > 1$ and Schur-convex if $0 < \beta < 1$.

As mentioned in Example 3.46, we have in any case the strong positive dependence property of MTP_2.

Example 4.44. (Conditionally i.i.d. gamma lifetimes). Let Θ be a non-negative random variable with a prior distribution Π_0. Let $T_1, ..., T_n$ be conditionally i.i.d., given $\{\Theta = \theta\}$ ($\theta > 0$), with gamma density

$$g(t|\theta) = \frac{\theta^\alpha}{\Gamma(\alpha)} t^{\alpha-1} \exp\{-\theta t\}, t \geq 0$$

where α is a positive constant. $g(t|\theta)$ is log-concave if $\alpha > 1$ and it is log-convex if $\alpha < 1$.

The joint density of $T_1, ..., T_n$ is then Schur-concave if $\alpha > 1$ and Schur-convex if $\alpha < 1$.

In the following example we consider again the case of conditionally i.i.d lifetimes. In this case, however, the conditional survival functions $\overline{G}(s|\theta)$ are not positive over all $\mathbb{R}_+$.

Example 4.45. Let Θ be a non-negative random variable with a prior density π_0 and let T_1, T_2 be conditionally i.i.d., given $\{\Theta = \theta\}$ ($\theta > 0$), with a distribution which is uniform over the interval $[0, \theta]$:

$$g(t|\theta) = \frac{1}{\theta} \mathbf{1}_{\{0 \leq t \leq \theta\}}$$

The joint density of $f^{(2)}(t_1, t_2)$ of T_1, T_2 is then

$$f^{(2)}(t_1, t_2) = \int_0^\infty \frac{1}{\theta^2} \mathbf{1}_{\{\theta \geq \max(t_1, t_2)\}} \pi_0(\theta) \, d\theta = \int_{\max(t_1, t_2)}^\infty \frac{\pi_0(\theta)}{\theta^2} d\theta.$$

In order to make explicit computation easy, we consider the particular case when π_0 is a Pareto density with parameters a_0 and $0 < u_0 < \max(t_1, t_2)$:

$$\pi_0(\theta) = (\alpha - 1) \cdot u^{\alpha-1} \frac{1}{\theta^a} \mathbf{1}_{\{\theta \geq u\}}.$$

This yields

$$f^{(2)}(t_1, t_2) = \frac{(\alpha - 1) \cdot u^{\alpha-1}}{(\alpha + 1) \left[\max(t_1, t_2)\right]^{\alpha+1}}.$$

which is Schur-concave, since it is a decreasing function of $\max(t_1, t_2)$.

Example 4.46. Consider the case of lifetimes $T_1, ..., T_n$ satisfying the condition N of negative dependence (Definition 3.56): there exists $s > 0$ and log-concave densities $g_1, ..., g_{n+1}$ such that the joint density of $T_1, ..., T_n$ is equal to the conditional density of $Z_1, ..., Z_{n+1}$ given $\sum_{i=1}^n Z_1 = s$, where $Z_1, ..., Z_{n+1}$ are independent variables with densities $g_1, ..., g_{n+1}$, respectively.

$T_1, ..., T_n$ are exchangeable if $g_1 = ... = g_{n+1} = g$ and the corresponding density is Schur-concave by Proposition 4.40.

The following examples show cases of Schur densities which have not been obtained from conditionally i.i.d. lifetimes or from conditioning i.i.d. lifetimes with respect to the sum.

Example 4.47. (Linear breakdown model). In the special Ross model with m.c.h.r. functions of the form

$$\lambda_t^{(n-h)}(t_1, ..., t_h) = \frac{1}{n-h},$$ (4.16)

we have the joint density

$$f^{(n)}(t_1, ..., t_n) = \frac{\theta^n}{n!} \cdot \exp\{-\theta \cdot \max(t_1, ...t_n)\}$$

which, by the same argument as above, is also Schur-concave. The interest of this example lies in that it is a case of Schur-concave density, with positive dependence, which has not been obtained as one of conditionally i.i.d. lifetimes.

Example 4.48. Consider the case of dependence, for a pair of lifetimes T_1, T_2, characterized by m.c.h.r. functions as follows

$$\lambda_t^{(2)} = \frac{\theta + a}{2}, \lambda_t^{(1)}(t_1) = \theta$$

where θ and a are given positive quantities.

The corresponding joint density is

$$f^{(2)}(t_1, t_2) = \theta \frac{\theta + a}{2} \exp\{-\theta \max(t_1, t_2) - a \min(t_1, t_2)\}.$$

If $a < \theta$, $f^{(2)}(t_1, t_2)$ is Schur-concave and we have a case of positive dependence; if $a > \theta$, $f^{(2)}(t_1, t_2)$ is Schur-convex and we have a case of negative dependence.

4.3.3   Properties of Schur densities

Propositions 4.39 and 4.40 show closure properties of the family of Schur-concave or Schur-convex density functions. The following results show further useful closure properties.

Proposition 4.49. *Let $f^{(n)}$ be a Schur-concave (Schur-convex) density function. Then its k-dimensional marginal $f^{(k)}$ is Schur-concave (Schur-convex) as well.*

Proof. For an arbitrary vector $(\xi_1, ..., \xi_{n-k}) \in \mathbb{R}_+^{n-k}$ we have the implication

$$(t'_1, ..., t'_k) \succ (t''_1, ..., t''_k) \Rightarrow$$

$$(t'_1, ..., t'_k, \xi_1, ..., \xi_{n-k}) \succ (t''_1, ..., t''_k, \xi_1, ..., \xi_{n-k}).$$

Then the assertion follows from the relation

$$f^{(k)}(t_1, ..., t_k) = \int_0^\infty \cdots \int_0^\infty f^{(n)}(t_1, ..., t_k, \xi_1, ..., \xi_{n-k}) \, d\xi_1 ... d\xi_{n-k}.$$

$\square$

As a corollary of Theorem 4.30, one obtains (see Marshall and Olkin, 1974)

Proposition 4.50. *Let* $\mathbf{T}_1$ *and* $\mathbf{T}_2$ *be two independent vectors of lifetimes with Schur concave joint densities* $f_1^{(n)}$ *and* $f_2^{(n)}$, *respectively. Then the vector* $\mathbf{T}_1 + \mathbf{T}_2$ *has a Schur-concave joint density.*

The assumption that the distribution of $\mathbf{T} \equiv (T_1, ..., T_n)$ admits a Schur density $f^{(n)}$ can be used to obtain some special properties for multivariate conditional hazard rates, normalized spacings, and TTT process. In particular one can obtain that some significant properties, valid in the case of vectors of i.i.d. lifetimes with monotone one-dimensional hazard rate functions, also hold for vectors of exchangeable lifetimes, under Schur conditions of the density functions.

Next, we shall show a result concerning the vector $\mathbf{C} \equiv (C_1, ..., C_n)$ of normalized spacings between order statistics (Definition 2.53), while in Subsection 4.4.1, we shall see a related result concerning the so called *TTT plot*.

From Section 2.4 we recall that the joint density of $\mathbf{C}$ is given by (2.74).

Remark 4.51. When $T_1, ..., T_n$ are i.i.d. and exponentially distributed, $C_1, ..., C_n$ are also i.i.d. and exponentially distributed. This is a special case of a remarkable property holding for any Schur constant joint density $f^{(n)}$ for $T_1, ..., T_n$.

Indeed from (2.74) we immediately obtain that, if $f^{(n)}$ is Schur-constant, then the joint density $f_{\mathbf{C}}$ of $C_1, ..., C_n$ coincides with $f^{(n)}$. In particular then, in such a case, $f_{\mathbf{C}}$ is Schur constant as well and $C_1, ..., C_n$ are exchangeable.

One can also see that, under regularity conditions for $f^{(n)}$, the condition $f_{\mathbf{C}} = f^{(n)}$ implies that $f_{\mathbf{C}}$ and $f^{(n)}$ are Schur-constant (see Ebrahimi and Spizzichino, 1997, for details).

Of course in general $C_1, ..., C_n$ cannot be exchangeable. Rather we can expect that, under conditions of aging, their one-dimensional marginal distributions are stochastically ordered.

In fact, in the case when $T_1, ..., T_n$ are i.i.d. with a one-dimensional survival function $\overline{F}$, the following result holds

Proposition 4.52. *(Barlow and Proschan, 1966, pg 1585). If* $\overline{F}$ *is* IFR (DFR), *then* $C_1 \geq_{st} \cdots \geq_{st} C_n$ $(C_1 \leq_{st} \cdots \leq_{st} C_n)$.

We shall prove that the same property holds for dependent lifetimes admitting Schur joint densities $f^{(n)}$. For this purpose we need the following lemma concerning a special aspect of the transformation $\mathcal{Z}$, introduced in Section 2.4.

For a vector $\mathbf{u} \equiv (u_1, ..., u_n) \in \mathbb{R}^n_+$ and for $i \in \{1, 2, ..., n-1\}$, denote by $\mathbf{M}_i(\mathbf{u})$ the vector

$$\mathbf{M}_i(\mathbf{u}) \equiv (u_1, ..., u_{i-1}, u_{i+1}, u_i, u_{i+2}, ..., u_n) \qquad (4.17)$$

obtained from $\mathbf{u}$ by interchanging the positions of the i-th coordinate and the $(i+1)$-coordinate.

The proof of the following Lemma is a bit tedious but elementary (see Exercise 4.76).

Lemma 4.53. *If* $u_i > u_{i+1}$, *then*

$$\mathcal{Z}(\mathbf{M}_i(\mathbf{u})) \succ \mathcal{Z}(\mathbf{u})$$

Proposition 4.54. *(see e.g. Ebrahimi and Spizzichino, 1997). If* $f^{(n)}$ *is Schur-concave (Schur-convex) then, for* $i = 1, ..., n-1$,

$$C_i \geq_{st} C_{i+1} \ (C_i \leq_{st} C_{i+1})$$

Proof. Let us consider the Schur-concave case. We must prove that, $\forall c > 0$, it is

$$P\{C_i > c\} \geq P\{C_{i+1} > c\}. \qquad (4.18)$$

Now

$$P\{C_i > c\} = \int_0^\infty d\xi_1 ... \int_0^\infty d\xi_{i-1} \int_c^\infty d\xi_i \int_0^\infty d\xi_{i+1} ... \int_0^\infty f_{\mathbf{C}}(\xi) \, d\xi_n \qquad (4.19)$$

By interchanging integration order, the r.h.s. of (4.19) can also be written

$$\int_0^\infty d\xi_1 ... \int_0^\infty d\xi_{i-1} \int_0^\infty d\xi_{i+2} ... \int_0^\infty d\xi_n \int_c^\infty d\xi_i \int_0^\infty f_{\mathbf{C}}(\xi) \, d\xi_{i+1}.$$

Similarly

$$P\{C_{i+1} > c\} = \int_0^\infty d\xi_1 ... \int_0^\infty d\xi_{i-1} \int_0^\infty d\xi_{i+2} ... \int_0^\infty d\xi_n \int_c^\infty d\xi_{i+1} \int_0^\infty f_{\mathbf{C}}(\xi) \, d\xi_i.$$

By writing

$$\int_c^\infty d\xi_i \int_0^\infty f_{\mathbf{C}}(\xi) \, d\xi_{i+1} = \int_c^\infty d\xi_i \left[\int_0^c f_{\mathbf{C}}(\xi) \, d\xi_{i+1} + \int_c^\infty f_{\mathbf{C}}(\xi) \, d\xi_{i+1} \right]$$

and

$$\int_c^\infty d\xi_{i+1} \int_0^\infty f_{\mathbf{C}}(\xi)\, d\xi_i = \int_c^\infty d\xi_{i+1} \left[\int_0^d f_{\mathbf{C}}(\xi)\, d\xi_i + \int_c^\infty f_{\mathbf{C}}(\xi)\, d\xi_i \right]$$

we see that, in order to achieve (4.18), we have only to show that, $\forall\,(\xi_1, ..., \xi_{i-1}, \xi_{i+2}, \cdot$

$$\int_c^\infty \int_0^c f_{\mathbf{C}}(\xi)\, d\xi_i d\xi_{i+1} \geq \int_c^\infty \int_0^c f_{\mathbf{C}}(\xi)\, d\xi_{i+1} d\xi_i$$

or, by taking into account Proposition 2.56,

$$\int_c^\infty \int_0^c f^{(n)}(\mathcal{Z}(\xi))\, d\xi_i d\xi_{i+1} \geq \int_c^\infty \int_0^c f^{(n)}(\mathcal{Z}(\xi))\, d\xi_{i+1} d\xi_i \qquad (4.20)$$

By using the notation introduced in (4.17), we can write

$$\int_c^\infty \int_0^c f^{(n)}(\mathcal{Z}(\xi))\, d\xi_{i+1} d\xi_i = \int_c^\infty \int_0^c f^{(n)}(\mathcal{Z}(\mathbf{M}_i(\xi)))\, d\xi_i d\xi_{i+1}$$

The validity of (4.20) can then be proved by taking into account that $f^{(n)}$ is Schur-concave and that, in the integration domain $\{\xi_i \geq c, 0 \leq \xi_{i+1} \leq c\}$, it is, by Lemma 4.53,

$$\mathcal{Z}(\mathbf{M}_i(\xi)) \succ \mathcal{Z}(\xi).$$

$\square$

Example 4.55. (Linear breakdown models). For the Ross model with joint density

$$f^{(n)}(\mathbf{t}) = \frac{\theta^n}{n!} \cdot \exp\{-\theta \max(t_1, ..., t_n)\}$$

one has

$$f_{\mathbf{C}}(\mathbf{c}) = \frac{\theta^n}{n!} \cdot \exp\left\{-\theta \cdot \sum_{i=1}^n \frac{c_i}{n-i+1}\right\}$$

i.e. $C_1, ..., C_n$ are independent and the distribution of C_i is exponential with mean $\mu_i = \frac{n-i+1}{\theta}$, $i = 1, ..., n$.

For a slightly more general example, see Exercise 4.73.

4.4   Further aspects of Bayesian aging

In this section we just sketch some other aspects which can be of interest for applications and further research. In the next subsection we concentrate attention on the notion of "scaled" TTT process.

We notice that, under Schur conditions for $f^{(n)}$ we can derive special properties for the random variables $Y_{(1)}, ..., Y_{(n)}$, defined in Section 2.4, by using Proposition 4.54.

In the statistical analysis of failure data, however, it is more convenient to look at "scaled" TTT process and at "scaled" total time on test until successive failures.

We shall then illustrate some properties which hold for such a notion in connection with aging properties of original lifetimes.

4.4.1   Schur densities and TTT plots

Denote as usual $S_n \equiv \sum_{i=1}^{n} T_i$.

Definition 4.56. The *scaled total time on test until h-th failure* is defined as the random quantity

$$Q_h \equiv \frac{Y_{(h)}}{S_n}, h = 1, ..., n - 1$$

For $h = n$, it is of course $Q_n \equiv \frac{Y_{(n)}}{S_n} = 1$.

The space of possible values of the random vector $\mathbf{Q} \equiv (Q_1, ..., Q_{n-1})$ is

$$\Gamma_{1,n-1} \equiv \left\{ \mathbf{t} \in \mathbb{R}_+^{n-1} | t_1 \leq t_2 \leq ... \leq t_{n-1} \leq 1 \right\}$$

and the joint density function of $\mathbf{Q}$ will be denoted by $g_{\mathbf{Q}}$. Schur properties of the joint density function of $(T_1, ..., T_n)$ are immediately reflected by $g_{\mathbf{Q}}$.

We have in fact

Proposition 4.57. *If $f^{(n)}$ is Schur-concave (Schur-convex) then $g_{\mathbf{Q}}$ is non-decreasing (non-increasing) on $\Gamma_{1,n-1}$, with respect to the natural partial ordering.*

We prefer to show applications of Proposition 4.57, before proceeding to provide a sketch of the proof. As a first application, we have

Corollary 4.58. *Let $f^{(n)}$ be an arbitrary Schur-constant density. Then the distribution of $\mathbf{Q}$ is the uniform distribution over $\Gamma_{1,n-1}$.*

Consider now two vectors of lifetimes $\mathbf{T} \equiv (T_1, ..., T_n)$ and $\mathbf{T}' \equiv (T_1', ..., T_n')$, admitting joint densities $f_{\mathbf{T}}^{(n)}$ and $f_{\mathbf{T}'}^{(n)}$ respectively.

As an interesting consequence of Proposition 4.57 and of the definition of multivariate $\leq_{lr}$ ordering (Definition 3.20) we have

Proposition 4.59. *If $f_{\mathbf{T}}^{(n)}$ is Schur-concave and $f_{\mathbf{T}'}^{(n)}$ is Schur-convex, then*

$$\mathbf{Q} \geq_{lr} \mathbf{Q}'.$$

Corollary 4.60. *If $f_{\mathbf{T}}^{(n)}$ is Schur-concave (or Schur-convex) and $f_{\mathbf{T}'}^{(n)}$ is Schur-constant, then*

$$\mathbf{Q} \geq_{lr} \mathbf{Q}' \qquad (or \ \mathbf{Q}' \geq_{lr} \mathbf{Q})$$

Remark 4.61. Consider lifetimes $\widehat{T}_1, ..., \widehat{T}_n$ with a Schur-constant density $f^{(n)}(\mathbf{t}) = \phi_n(\sum_{i=1}^{n} t_i)$.

The distribution of the corresponding $\widehat{\mathbf{Q}}$ is uniform over $\Gamma_{1,n-1}$ and it is independent of ϕ_n.

In other words, for all vectors having a Schur-constant density, the distribution of the corresponding scaled Total Times on Test is the same as in the case of i.i.d. $T_1, ..., T_n$ with a standard exponential distribution.

Corollary 4.60 may be seen as a counterpart, for the case of stochastic dependence, of a result valid for i.i.d. lifetimes. In fact, a well-known result is the following

Proposition 4.62. *(Barlow and Campo, 1975). Let $T_1, ..., T_n$ be i.i.d. with an IFRA (or, in particular IFR) distribution. Then, for the corresponding scaled total time on test, it is*

$$\mathbf{Q} \geq_{st} \widehat{\mathbf{Q}}.$$

This result can be used to obtain, under conditions of aging, useful properties of the *scaled total time on test plot* associated with the longitudinal observation of lifetimes $T_1, ..., T_n$. This is defined as the polygonal line obtained by joining the $(n + 1)$ points of coordinates

$$(0,0), \left(\frac{1}{n}, Q_1\right), \left(\frac{2}{n}, Q_2\right), ..., \left(\frac{n-1}{n}, Q_{n-1}\right), (1,1).$$

This statistic appears to be a very interesting one in the analysis of failure data and in particular in the derivation of burn-in and age replacement procedures (see in particular Barlow and Campo, 1975; Bergman, 1979; Cooke, 1993). In the case of i.i.d. lifetimes it can be seen as a sort of empirical counterpart of the notion of *total time on test transformation* (see Shorack, 1972; Bergman and Klefsjö, 1985; Chandra and Singpurwalla, 1981; Gill, 1986; Klefsjö, 1982).

We now give a sketch of the proof of Proposition 4.57; for details see Nappo and Spizzichino (1998). A first step is to derive the joint law of $(Q_1, ..., Q_{n-1})$.

To this purpose note that, by taking into account the definitions of $\left(Y_{(1)}, ..., Y_{(n)}\right)$ and $(C_1, ..., C_n)$, one can write

$$\left(T_{(1)}, ..., T_{(n)}\right) = \mathcal{T}\left(Q_1, ..., Q_{n-1}, S_n\right)$$

where $\mathcal{T}$ is the one-to-one differentiable transformation defined by

$$\mathcal{T}\left(Q_1, ..., Q_{n-1}, S_n\right) = \left(\rho_1(\mathbf{Q}) \cdot S_n, ..., \rho_n(\mathbf{Q}) \cdot S_n\right)$$

with

$$\rho_j(\mathbf{q}) = \frac{q_j}{n - j + 1} - \sum_{i=1}^{j-1} \left(\frac{1}{n - i + 1} - \frac{1}{n - 1}\right) q_i, j = 1, ..., n \qquad (4.21)$$

Note also that it is

$$\sum_{j=1}^{n} \rho_j(\mathbf{q}) = 1, \forall \mathbf{q} \equiv (q_1, ..., q_{n-1}) \in \Gamma_{1,n-1} \qquad (4.22)$$

By some computation it can be shown that the Jacobian of the transformation $\mathcal{T}$ is $\left|J_{\mathcal{T}}\left(q_1, ..., q_{n-1}, s\right)\right| = \frac{s^{n-1}}{n!}$.

By recalling Lemma 2.55, we then obtain

Proposition 4.63. *The vector* $(Q_1, ..., Q_{n-1}, S_n)$ *has the joint density*

$$g_{(Q_1, ..., Q_{n-1}, S_n)}\left(q_1, ..., q_{n-1}, s\right) =$$

$$f^{(n)}\left(\rho_1(\mathbf{q}) \cdot s, \rho_2(\mathbf{q}) \cdot s, ..., \rho_n(\mathbf{q}) \cdot s\right) \cdot s^{n-1}$$

Whence

Corollary 4.64. *The vector* $\mathbf{Q}$ *has the joint density*

$$g_{\mathbf{Q}}\left(q_1, ..., q_{n-1}\right) =$$

$$\mathbf{1}_{\Gamma_{1,n-1}}\left(q_1, ..., q_{n-1}\right) \cdot \int_0^\infty f^{(n)}\left(\rho_1(\mathbf{q}) \cdot s, ..., \rho_n(\mathbf{q}) \cdot s\right) \cdot s^{n-1} ds$$

Proposition 4.57 can finally be obtained from the following property of the transformation $\mathcal{T}$:

Lemma 4.65. *Let* $\mathbf{q}, \mathbf{q}' \in \Gamma_{1,n-1}$. *If* $q_i \geq q_i'$ *for* $i = 1, ..., n - 1$, *then*

$$\left(\rho_1(\mathbf{q}'), \rho_2(\mathbf{q}'), ..., \rho_n(\mathbf{q}')\right) \succ \left(\rho_1(\mathbf{q}), \rho_2(\mathbf{q}), ..., \rho_n(\mathbf{q})\right).$$

4.4.2   Some other notions of Bayesian aging

As said in the first section of this chapter, one can define several notions of aging for exchangeable lifetimes, by fixing attention on a given univariate notion and then extending it to the exchangeable case, by requiring the validity of an inequality of the type in (4.3).

Following this line, we discussed two different multivariate extensions of the notion of IFR, one of which is equivalent to Schur-concavity of the joint survival function and the other, valid for the absolutely continuous case, is implied by Schur-concavity of the joint density.

Here we mention a few further definitions along the same direction. We start with the multivariate extension of the notion of NBU. The univariate notion for i.i.d. lifetimes is, in particular, equivalent to

$$P\{T_1 > \tau|D\} \geq P\{T_2 > s_2 + \tau|D\}, \forall \tau \tag{4.23}$$

where

$$D \equiv \{(T_2 > s_2) \cap H\} \tag{4.24}$$

with

$$H \equiv \{T_3 > s_3, ..., T_h > s_n\}. \tag{4.25}$$

It is immediately obvious that (4.23) is equivalent to

$$\overline{F}^{(n)}(\tau, s_2, s_3, ..., s_n) \geq \overline{F}^{(n)}(0, s_2 + \tau, s_3, ..., s_n) = \overline{F}^{(n-1)}(s_2 + \tau, s_3, ..., s_n) \tag{4.26}$$

and that it is implied by Schur-concavity of $\overline{F}^{(n)}$.

The condition (4.26) can then be seen as a multivariate definition of NBU for exchangeable lifetimes. Some aspects of this definition have been analyzed by Bassan and Spizzichino (2000). In particular, it is easy to see that, when $T_1, ..., T_n$ are conditionally i.i.d., with NBU distribution, given a parameter Θ, then (4.23) holds for the unconditional survival function (see Exercise 4.78).

Let us now come to multivariate notions related to stochastic comparisons stronger than $\geq_{st}$. Here we just mention a couple of cases, focusing attention on situations of positive aging. A more complete treatment, with additional cases, examples and proofs is presented in Bassan and Spizzichino (1999).

As seen in Section 3.4, a univariate condition, stronger than IFR, is that a lifetime T is PF_2, i.e. that the density f of T is log-concave, which is equivalent to the stochastic comparison in (3.22).

A multivariate extension to exchangeable lifetimes $T_1, ..., T_n$ is then provided by the condition

$$\mathcal{L}(T_1 - s_1|D) \geq_{lr} \mathcal{L}(T_2 - s_2|D), \text{ for } 0 \leq s_1 < s_2. \tag{4.27}$$

where

$$D \equiv \{(T_1 > s_1) \cap (T_2 > s_2) \cap H\}$$

with H containing survival and failure data for lifetimes $T_3, ..., T_n$. A set of sufficient conditions for (4.27) is:

(i) $f^{(n)}$ is log-concave

(ii) $f^{(n)}$ is MTP$_2$.

An example is presented in Exercise 4.81.

We also saw that, in the univariate case, the IFR property is equivalent to the apparently stronger condition

$$\mathcal{L}(T - s_1 | T > s_1) \geq_{hr} \mathcal{L}(T - s_2 | T > s_2), \forall 0 \leq s_1 < s_2$$

In other words this means that, for i.i.d. lifetimes $T_1, ... T_n$ the two conditions

$$\mathcal{L}(T_1 - s_1 | D) \geq_{st} \mathcal{L}(T_2 - s_2 | D), \forall 0 \leq s_1 < s_2 \qquad (4.28)$$

$$\mathcal{L}(T_1 - s_1 | D) \geq_{hr} \mathcal{L}(T_2 - s_2 | D), \forall 0 \leq s_1 < s_2 \qquad (4.29)$$

coincide for $D \equiv \{(T_1 > s_1) \cap (T_2 > s_2) \cap ... \cap (T_n > s_n)\}$.

This is not anymore true in general when $T_1, ... T_n$ are exchangeable but stochastically dependent. A set of sufficient conditions for (4.29) is:

(i) The mapping $s_1 \to \overline{F}^{(n)}(s_1, s_2, s_3, ..., s_n)$ is log-concave

(ii) The mapping $(s_1, s_2) \to \overline{F}^{(n)}(s_1, s_2, s_3, ..., s_n)$ is TP$_2$.

See Exercises 4.79, 4.80, relative to the case $n = 2$.

4.4.3   Heterogeneity and multivariate negative aging

It is well known that, for one-dimensional lifetime distributions, situations of heterogeneity create a tendency to negative aging. This is shown in particular by Remark 3.66 and by Proposition 3.65 which states that a mixture of DFR distributions is DFR (then in particular a mixture of exponential distributions is DFR). More in general it has been noticed that properties of negative aging can even arise in the case of mixtures of non-necessarily DFR distributions.

In this subsection we generalize the exchangeable hierarchical model of heterogeneous populations, considered in several examples in the previous chapters. The model we obtain accounts, at a time, for heterogeneity and for interdependence among lifetimes; it allows us to clarify some facts related to heterogeneity and aging; in particular we shall see multivariate analogs of Proposition 3.65 and Remark 3.66.

Consider a population formed of n individuals $U_1, ..., U_n$ and let (K_i, T_i) be a pair of random variables corresponding to U_i, for $i = 1, ..., n$. We think of

T_i as of an observable lifetime, while Z_i is an unobservable (endogenous to U_i) quantity which influences the distribution of T_i (and only of T_i).

$K_1, ..., K_n$ are random variables taking their values in a set $\mathcal{K} \subset \mathbb{R}$ (in the previous examples of exchangeable heterogeneous populations the special case $\mathcal{K} \equiv \{0, 1\}$ was considered).

Then the marginal distribution of T_i is a mixture (the mixture of conditional distributions given $(K_i = k)$, $k \in \mathcal{K}$).

Put $\mathbf{K} \equiv (K_1, ..., K_n)$ and assume, more precisely, the existence of a family $\{\overline{G}(t|k), k \in \mathcal{K}\}$ of one-dimensional survival functions such that the conditional survival function of T_i, given $\{\mathbf{K} = \mathbf{k}\}$ is $\overline{G}(t|k_i)$.

Assuming $T_1, ..., T_n$ to be conditionally independent given $\mathbf{K}$, we see that the joint conditional survival function of $T_1, ..., T_n$ given $\mathbf{K}$ is

$$\overline{F}^{(n)}(t_1, ..., t_n | k_1, ..., k_n) = \prod_{i=1}^{n} \overline{G}(t_i|k_i) \tag{4.30}$$

so that, by denoting $\pi_0(\mathbf{k})$ the joint density of $\mathbf{K}$, the unconditional survival function of $T_1, ..., T_n$ is

$$\overline{F}^{(n)}(t_1, ..., t_n) = \int_{\mathcal{K} \times ... \times \mathcal{K}} \prod_{i=1}^{n} \overline{G}(t_i|k_i)\pi_0(\mathbf{k})dk_1, ..., dk_n \tag{4.31}$$

Let us first consider the case when $K_1, ..., K_n$ are i.i.d. and denote by p_0 their common density, so that their joint density is

$$\pi_0(k_1, ..., k_n) = \prod_{i=1}^{n} p_0(k_i). \tag{4.32}$$

In such a case the unconditional survival function of $T_1, ..., T_n$ is

$$\overline{F}^{(n)}(t_1, ..., t_n) = \int_{\mathcal{K} \times ... \times \mathcal{K}} \prod_{i=1}^{n} \overline{G}(t_i|k_i) \prod_{i=1}^{n} p_0(k_i)dk_1, ..., dk_n \tag{4.33}$$

and it is clear that $T_1, ..., T_n$ are i.i.d. as well, their common one-dimensional survival function being

$$\overline{F}^{(1)}(t) = \int_{\mathcal{K}} \overline{G}(t|k)p_0(k)dk. \tag{4.34}$$

Recalling the Corollary 4.13, we reformulate Proposition 3.65 and Remark 3.66, respectively, as follows:

Proposition 4.66. *If* $\overline{G}(t|k)$ *is DFR,* $\forall k \in \mathcal{K}$, *then* $\overline{F}^{(n)}(t_1, ..., t_n)$ *in (4.33) is Schur-convex.*

Remark 4.67. $\overline{G}(t|k)$ *being IFR* $\forall k \in \mathcal{K}$, *does not necessarily imply that* $\overline{F}^{(n)}(t_1, ..., t_n)$ *in (4.33) is Schur-concave.*

Here, we are interested in the case when $K_1, ..., K_n$ are not independent and then when also $T_1, ..., T_n$ are interdependent. More specifically, we consider the case when the initial distribution of $\mathbf{K} \equiv (K_1, ..., K_n)$ has an exchangeable density function $\pi_0^{(n)}(\mathbf{k})$. In such a case also $T_1, ..., T_n$ are exchangeable.

Proposition 3.65 and Remark 3.66, respectively, can be extended to this case as follows.

Remark 4.68. $\overline{G}(t|k)$ being IFR $\forall k \in \mathcal{Z}$, does not necessarily imply that $\overline{F}^{(n)}(t_1, ..., t_n)$ is Schur-concave.

The difference between this claim and the one in Remark 4.67 is just in the fact that there we considered survival functions of the form (4.33), while here the case (4.31) is considered.

Proposition 4.69. Let $\frac{\overline{G}(t|k)}{\overline{G}(t|k')}$ be non-decreasing in t, for $k < k'$. If $\overline{G}(t|k)$ is DFR $\forall k \in \mathcal{K}$, then $\overline{F}^{(n)}(t_1, ..., t_n)$ in (4.31) is Schur-convex for any exchangeable joint density $\pi_0^{(n)}(\mathbf{k})$.

Note that the condition that $\frac{\overline{G}(t|k)}{\overline{G}(t|k')}$ is non-decreasing in t, for $k < k'$ means that K_i can be seen as a "frailty" for the individual U_i.

The proof of Proposition 4.69 substantially relies on the following facts:

letting D be an event as in (4.2), we have, by the above assumption of conditional independence,

$$P\{T_i > s_i + \tau | D\} = \mathbb{E}[\frac{\overline{G}(s_i + \tau | K_i)}{\overline{G}(s_i | K_i)} | D]; \qquad (4.35)$$

for $0 \leq s_1 < s_2$,

$$P\{K_1 > k | D\} \geq P\{K_2 > k | D\}, \forall k \in \mathcal{K}.$$

A complete proof is contained in Spizzichino and Torrisi (2001). Such a result therein is obtained as a corollary of a more general result, concerning a multivariate extension of the concept of *ultimately DFR* univariate distributions, which was considered for instance in Block, Mi and Savits (1993). Simple examples can be found, which confirm what was claimed in Remark 4.68. First examples are in particular provided by the case $K_1, ..., K_n$ i.i.d., when indeed Remark 4.68 reduces to Remark 3.66. However in the extreme case when

$$P\{K_1 = K_2 ... = K_n\} = 1 \qquad (4.36)$$

then $T_1, ..., T_n$ are conditionally independent, identically distributed given K_1 and, assuming $\overline{G}(t|k)$ IFR $\forall k \in \mathcal{K}$, we actually obtain that $\overline{F}^{(n)}(t_1, ..., t_n)$ is Schur-concave, since now we reduce to the case of Proposition 4.18.

This agrees with the fact that, in such an extreme case of positive dependence among $K_1, ..., K_n$, the situation of heterogeneity disappears.

In this respect, the case is particularly interesting when

$$\overline{G}(t|k) = \exp\{-tk\}. \tag{4.37}$$

Under this choice, $\overline{F}^{(n)}(t_1, ..., t_n)$ is Schur-constant if (4.36) holds, while we have Schur-convexity in the strict sense if the condition (4.36) is replaced by the opposite condition $K_1, ..., K_n$ i.i.d. (with a nondegenerate distribution).

Further aspects of the model of heterogeneity described above are studied in Gerardi, Spizzichino and Torti (2000a), where a study is carried out of the case when $K_1, ..., K_n$ are discrete random variables, and in Gerardi, Spizzichino and Torti (2000b), where the conditional distribution of residual lifetimes, given a dynamic history, is studied as a stochastic filtering problem.

4.4.4 A few bibliographical remarks

Some general aspects of probability distributions with Schur properties were studied in the paper by Marshall and Olkin (1974).

As mentioned a complete treatment about majorization and Schur-concavity (Schur-convexity) is presented in the book by Marshall and Olkin (1979).

In the specific field of reliability the concept of majorization was used by Pledger and Proschan (1971), by Proschan (1975), and by several other Authors later on, as accounted in the review paper by Boland and Proschan (1988).

Starting in the late eighties, interest about notions of multivariate aging, suitable for a Bayesian approach to reliability, started to arise among a number of different people.

This was mainly related to research on the meaning of infant mortality (in the case of interdependence) and to the study of optimality of burn-in.

In particular there were several discussions between Richard E. Barlow and the author.

The idea that Schur properties of joint densities of lifetimes is a notion of multivariate aging is contained in Barlow and Mendel (1992). A discussion on that was presented by R. E. Barlow in the Fall of 1990 at a Conference on Reliability and Decision Making, held in Siena (Italy); it came out rather naturally from a comparison with the particular Schur-constant case, which conveys the idea of *indifference with respect to age,* or of a Bayesian, multivariate, counterpart of exponentiality.

4.4.5 Extensions to non-exchangeable cases

In many cases, properties of aging for a vector of lifetimes are interesting in view of the fact that they can be converted into corresponding probabilistic assessments on the behavior of the vector of order statistics (see for instance Subsection 4.1 above).

Here we considered notions of aging for exchangeable lifetimes; however it is well known that, for an arbitrary vector of random variables $X_1, ..., X_n$, we can build a vector of exchangeable random variables $T_1, ..., T_n$ in such a way that the two corresponding vectors of order statistics have the same law; this is in particular true by letting

$$T_i \equiv X_{\mathrm{P}_i}, i = 1, ..., n$$

where P is a random permutation of $\{1, 2, ..., n\}$. If $f_{\mathbf{X}}$ is the joint density of $\mathbf{X}$, then the joint density $f_{\mathbf{T}}$ of $\mathbf{T}$ is the exchangeable density obtained by *symmetrization*:

$$f_{\mathbf{T}}(\mathbf{t}) = \frac{1}{n!} \sum_{\pi} f_{\mathbf{X}}(t_{\pi_1}, ..., t_{\pi_n}) \tag{4.38}$$

where the sum is extended to all the permutations of $\{1, 2, ..., n\}$.

Then a property of aging for $f_{\mathbf{T}}$ in (4.38) can, in a sense, be seen as a property of aging for $f_{\mathbf{X}}$.

This shows that it can be of interest to find sufficient conditions on $f_{\mathbf{X}}$ which guarantee a given property of aging for $f_{\mathbf{T}}$.

A result of this kind (see Marshall and Olkin, 1979, p. 83) is provided by the following.

Proposition 4.70. *If $f_{\mathbf{X}}$ is convex (but not necessarily permutation-invariant) then $f_{\mathbf{T}}$ is Schur-convex.*

This result is also used in the proof of Theorem 3.1 in Kochar and Kirmani (1995).

4.5   Exercises

Exercise 4.71. Consider the joint survival function $\overline{F}^{(n)}$ for the model of heterogeneity described in Examples 2.3 and 2.20. Assume that the failure rate functions $r_0(t)$ and $r_1(t)$ are non-increasing and such that $r_0(t) \leq r_1(t)$ and that $K_1, ..., K_n$ are (exchangeable) binary random variables.

Show that $\overline{F}^{(n)}$ is Schur-convex.

Hint: Show that, for $D \equiv \{T_1 > s_1, ..., T_n > s_n\}$, $0 \leq s_1 \leq s_2$, and arbitrary $s_3, ..., s_n \leq 0$, it is

$$P\{K_1 = 1|D\} \geq P\{K_2 = 1|D\}.$$

Then take into account conditional independence of $T_1, ..., T_n$ given $K_1, ..., K_n$ and arguments in Remark 4.16 (see also arguments in Subsection 4.4).

Exercise 4.72. Consider the change-point model in Example 2.19 and assume

$$r_1(t) = \theta_1, \quad r_2(t) = \theta_2, \quad q(\sigma) = \mu \exp\{-\mu\sigma\}.$$

Find an expression for $\overline{F}^{(2)}$ and, by applying the Schur's condition (4.8), check if $\overline{F}^{(2)}$ is Schur-concave.

Hint: for $s_1 \leq s_2$,

$$\overline{F}^{(2)}(s_1, s_2) = \int_0^{s_1} \overline{F}^{(2)}(s_1, s_2|\sigma) q(\sigma) d\sigma +$$

$$\int_{s_1}^{s_2} \overline{F}^{(2)}(s_1, s_2|\sigma) q(\sigma) d\sigma \int_{s_2}^{\infty} \overline{F}^{(2)}(s_1, s_2|\sigma) q(\sigma) d\sigma.$$

Exercise 4.73. Show that, for a Ross model characterized by the condition

$$\lambda_t^{(n-h)}(t_1, ..., t_h) = \varphi_h, \lambda_t^{(n)} = \varphi_0,$$

with $\varphi_0 \leq \varphi_1 \leq ... \leq \varphi_{n-1}$, the MIFR property holds and

$$C_1 \geq_{st} \cdots \geq_{st} C_n.$$

Hint: Take into account Definition 2.61 and Definition 3.70.

Exercise 4.74. Consider a joint density of the form

$$f^{(n)}(\mathbf{t}) = \psi_n \left(\max_{1 \leq i \leq n} t_i \right).$$

Show that ψ_n being non-increasing implies, for the normalized spacings between order statistics,

$$C_1 \geq_{st} \cdots \geq_{st} C_n$$

and is equivalent to the MIFR property of $f^{(n)}$.

Exercise 4.75. (Schur-constant, non-infinitely extendible models). $T_1, ..., T_n$ and $Y_1, ..., Y_m$ are the lifetimes of individuals forming two different populations $\mathcal{P}_1$ and $\mathcal{P}_2$, respectively. The individuals in the two populations are competing in the following sense: they need the same resource A for their lives. In particular the need of A for the individual j in $\mathcal{P}_1$ is proportional to the lifetime T_j $(j = 1, ..., n)$. The initial amount of A is $a > 0$. The individuals in $\mathcal{P}_1$ use the resource A before the individuals in $\mathcal{P}_2$ do. If $\{T_1 = t_1, ..., T_n = t_n\}$, the total amount of A left for individuals in $\mathcal{P}_2$ is $a - \sum_{j=1}^{n} t_j$. This amount is shared among different individuals, and each of them has the same distribution with mean $\frac{a - \sum_{j=1}^{n} t_j}{m}$. Assume that the initial density of $T_1, ..., T_n$ is Schur-constant.

Show that, conditionally on $\{Y_1 = y_1, ..., Y_m = y_m\}$, the density of $T_1, ..., T_n$ is Schur-constant and that $T_1, ..., T_n$ are not i.i.d. or conditionally i.i.d.

Exercise 4.76. Prove Lemma 4.53.

Hint: Notice that, $\forall \mathbf{a} \in \mathbb{R}_+^n$ the components of the vector $\mathbf{u} \equiv \mathcal{Z}(\mathbf{a})$ are ordered in the increasing sense:

$$u_1 \leq \dots \leq u_n,$$

i.e. $u_i = u_{(i)}$, $i = 1, \dots, n$.

Exercise 4.77. Prove Lemma 4.65.

Exercise 4.78. Check that, for lifetimes $T_1, \dots, T_n$, conditionally i.i.d. with conditional NBU distributions given a parameter Θ, the inequality in (4.26) holds.

Exercise 4.79. Let (T_1, T_2) be exchangeable lifetimes with survival function $\overline{F}^{(2)}$ satisfying:

(a) $\overline{F}^{(2)}$ is TP_2

(b) the mapping $s_1 \to \overline{F}^{(2)}(s_1, s_2)$ is log-concave.
For $D \equiv \{T_1 > r_1, T_s > r_2\}$, $0 \leq r_1, r_2$, prove that

$$\mathcal{L}(T_1 - r_1 | D) \geq_{hr} \mathcal{L}(T_2 - r_2 | D)$$

i.e. that

$$\mathcal{R}(t) \equiv \frac{P\{T_1 - r_1 > t | D\}}{P\{T_2 - r_2 > t | D\}}$$

is increasing in t.

Exercise 4.80. (Bivariate models of Marshall-Olkin). Check that, for the joint survival function

$$\overline{F}^{(2)}(s_1, s_2) = \exp\{-\lambda(t_1 + t_2) - \lambda' \max(t_1, t_2)\}$$

the conditions (a) and (b) of Exercise 4.79 hold.

Exercise 4.81. Consider again the model with joint density of the form

$$f^{(n)}(\mathbf{t}) = \frac{\theta^n}{n!} \exp\{-\theta \max_{1 \leq i \leq n} t_i\}.$$

Show that the inequality (4.27) holds.

4.6 Bibliography

Barlow, R.E. and Campo, R. (1975). Total time on test processes and applications to failure data analysis. In *Reliability and fault tree analysis*, R.E. Barlow, J. Fussel and N. Singpurwalla Eds., SIAM, Philadelphia, 451-481.

Barlow, R.E. and Proschan, F. (1966). Inequalities for linear combinations of order statistics from restricted families. *Ann. Math. Statist.*, 37, 1574-1592.

Barlow, R.E. and Proschan, F. (1975). *Statistical theory of reliability and life-testing.* Probability models. Holt, Rinehart and Winston, New York.

Barlow, R.E. and Mendel, M.B. (1992). de Finetti-Type representations for life-distributions. *J. Am. Statis. Soc.*, 87, 1116-1122.

R.E. Barlow and Mendel, M.B. (1993). Similarity as a characteristic of wear-out. In *Reliability and decision making*, R.E. Barlow, C.A. Clarotti and F. Spizzichino, Eds., Chapman & Hall, London.

Barlow, R.E. and Spizzichino, F. (1993). Schur-concave survival functions and survival analysis. *J. Comp. Appl. Math.*, 46, 437-447.

Bartoszewicz, J. (1986). Dispersive ordering and the total time on test transformation. *Statist. Probab. Lett.*, 4, 285-288.

Bassan, B. and Spizzichino, F. (1999). Stochastic comparisons for residual lifetimes and Bayesian notions of multivariate ageing. *Adv. Appl. Probab.* 31, no. 4, 1078-1094.

Bassan, B. and Spizzichino, F. (2000). On a multivariate notion of New Better than Used. *Volume of contributed papers presented at the Conference "Mathematical Methods of Reliability", Bordeaux, July 2000.*

Bergman, B. (1979). On age replacement and the total time on test concept. *Scand. J. Statist.*, 6, 161-168.

Bergman, B. and Klefsjö,B. (1985). Burn-in models and TTT-Transforms. *Quality and Reliability Engineering International*, 1, 125-130.

Block, H.W., Mi, J. and Savits, T.H. (1993). Burn in and mixed populations. *J. Appl. Probab.*, 30, 692-702.

Boland, P.J. and Proschan, F. (1988). The impact of reliability theory on some branches of mathematics and statistics. In *Handbook of Statistics*, Vol. 7, P.R. Krishnaiah and C.R. Rao Eds, Elsevier Science Pub., 157-174.

E.C. Brindley and W.A. Thompson (1972) Dependence and Aging Aspects of Multivariate Survival. *JASA*, Vol. 67, pg. 822-830.

Caramellino, L. and Spizzichino, F. (1994). Dependence and Aging Properties of Lifetimes with Schur-constant Survival Functions. *Prob. Engrg. Inform. Sci.*, 8, 103-111.

Chandra, M. and Singpurwalla, N. (1981). Relationships between some notions which are common to reliability theory and Economics. *Math. Op. Res.*, 6, 113-121.

Chick, S. E. and Mendel, M. B. (1998). New characterizations of the no-aging property and the l_1−isotropic model. *J. Appl. Probab.*, 35, no. 4, 903–910

Cooke, R. (1993). The total time on test statistic and age-dependent censoring. *Statist. Probab. Lett.*, 18, 307-312.

Ebrahimi, N. and Spizzichino, F. (1997). Some results on normalized total time on test and spacings. *Statist. Probab. Lett.* 36 (1997), no. 3, 231–243.

Gerardi, A., Spizzichino, F. and Torti, B. (2000a). Exchangeable mixture models for lifetimes: the role of "occupation numbers". *Statist. Probab. Lett.*, 49, 365-375.

Gerardi A., Spizzichino, F. and Torti, B. (2000b). Filtering equations for the conditional law of residual lifetimes from a heterogeneous population. *J. Appl. Probab.*, 37, no. 3, 823–834.

Gill, R.D. (1986). The total time on test plot and the cumulative total time on test statistics for a counting process. *Ann. Statist.*, 4, 1234-1239.

Hayakawa, Y. (1993). Interrelationships between l_p-isotropic densities and l_p-isotropic survival functions, and de Finetti representations of Schur-concave survival functions. *Austral. J. Statist.* 35, no. 3, 327–332.

Kochar, S. C. and Kirmani, S. N. U. A. (1995). Some results on normalized spacings from restricted families of distributions. *J. Statist. Plann. Inference* 46, no. 1, 47–57.

Klefsjö, B. (1982). On aging properties and total time on test transforms. *Scand. J. Statist.*, 9, 37-41.

Langberg, N. A., Leon, R.V., and Proschan, F. (1980). Characterizations of nonparametric classes of life distributions. *Ann. Probab.*, 8, 1163-1170.

Marshall, A.W. and Olkin I. (1974). Majorization in multivariate distributions. *Ann. Statist.*, 2, 1189-1200.

Marshall, A.W. and Olkin I. (1979). *Inequalities: theory of majorization and its applications.* Academic Press, New York.

Nappo, G. and Spizzichino, F. (1998). Ordering properties of the TTT-plot of lifetimes with Schur joint densities. *Statist. Probab. Lett.* 39, no. 3, 195–203.

Ostrowski, A. M. (1952). Sur quelques applications des fonctions convexes et concaves au sens de I. Schur. *J. Math. Pures Appl.*, 31, 253-292.

Pledger G. and Proschan, F. (1971). Comparisons of order statistics and of spacings from heterogeneous distributions. *Optimizing methods in statistics.* Academic Press, New York, 89–113.

Proschan, F. (1975). Applications of majorization and Schur functions in reliability and life testing. In *Reliability and fault tree analysis*, Soc. Indust. Appl. Math., Philadelphia, Pa, 237–258.

Schur, I. (1923). Uber eine Klasse von Mittelbildungen mit Anwendungen die Determinanten-Theorie Sitzungber. *Berlin Math. Gesellshaft*, 22, 9-20.

Shorack, G. (1972). Convergence of quantile and spacing processes with applications. *Ann. Math. Statist.*, 43, 1400-1411.

Spizzichino, F. (1992). Reliability decision problems under conditions of ageing. In *Bayesian Statistic 4*, J. Bernardo, J. Berger, A.P. Dawid, and A.F.M. Smith Eds., Clarendon Press, Oxford, 803-811.

Spizzichino, F. and Torrisi, G. (2001). Multivariate negative aging in a exchangeable model of heterogeneity. *Statist. Probab. Lett.* (To appear).

Chapter 5

Bayesian decisions, orderings, and majorization

5.1 Introduction

In this chapter we discuss some aspects of Bayesian decision problems arising in the field of reliability. We shall also see the role that notions of stochastic orderings, dependence, and majorization can have in those problems. We shall in particular concentrate attention on the cases of two-action decision problems and of burn-in problems. Some aspects of two-action decision problems and of burn-in problems will be discusssed in Section 5.2 and Section 5.4, respectively. Section 5.3 will be specifically devoted to analyzing the possible role of the notion of majorization, when comparing the informational effects of two different sets of survival/failure data, with the same value for the total-time-on-test statistic.

In this section, the language and some basic facts concerning *decision problems under uncertainty* will be briefly recalled; moreover we shall give the shape of decision problems to a few classical problems of the theory of reliability. For further discussions and examples of the role of stochastic orderings in decision problems and reliability see e.g. Torgersen (1994); Block and Savits (1994).

A Bayes decision problem under uncertainty is specified by the elements

$$A, \mathcal{V}, l, \Pi$$

where
> A is *the space of possible "actions" (or "decisions")*
> $\mathcal{V}$ is *the space of possible values for an unobservable quantity W*
> Π is *the probability distribution of W*
> $l: A \times \mathcal{V} \to \mathbb{R}$ is *the loss function*

In words the problem amounts to determining an optimal action $a^* \in A$, taking into account that the consequence of an action a depends on the value taken by W, which is unknown to us, at least at the instant when the decision is to be taken.

The loss, coming from the choice of a when the value of W is w, is measured by the scalar quantity $l(a; w)$ and our uncertainty on W is specified by means of the distribution Π.

In the statisticians' language, W is sometimes called *"state of nature"*.

Commonly the loss $l(a; w)$ is thought of as the opposite of the "utility" coming from the consequence of choosing a if the value of W were w.

This motivates that we can consider as an optimal action the one which minimizes the expected value of the loss, namely we consider the following definition of optimality, corresponding to the principle of maximization of expected utility (see e.g. Savage, 1972, De Groot, 1970, Lindley, 1985, Berger, 1985).

When we fix a loss function $l(a; w)$, we consider the loss $l(a; W)$ for any action $a \in A$.

W being a random variable, $l(a; W)$ is itself random (see the brief discussion in Subsection 5.1.3); the probability distribution of $l(a; W)$ depends on l and on the distribution of W. When the distribution of W is Π, the expected value of $l(a; W)$ is given by

$$\mathcal{R}_\Pi(a) \equiv \mathbb{E}_\Pi [l(a; W)] = \int_{\mathcal{V}} l(a; w) d\Pi(w)$$

$\mathcal{R}_\Pi(a)$ is the "risk of a against Π".

Definition 5.1. For a fixed loss function l, a_Π^* is a Bayes (or optimal) action against the distribution Π, if

$$\mathcal{R}_\Pi(a_\Pi^*) \leq \mathcal{R}_\Pi(a) \tag{5.1}$$

for all $a \in A$.

Of course, when A is not a finite space, it can happen that no Bayes action exists.

Example 5.2. (A typical two-action problem). For a given unit U, we have to decide whether it is good enough for a pre-established task; say for instance that the latter requires a mission time τ.

This can be seen as a decision problem in which $A \equiv \{a_1, a_2\}$ with:

$a_1 = \{$discarding the unit $U\}$; $a_2 = \{$using U for the mission$\}$.

The unobservable quantity W of interest here obviously coincides with T, the lifetime of U.

For a given loss function $l(a,t)$ $(a \in A, t \geq 0)$ the optimal decision will be a_2 if and only if

$$\mathbb{E}[l(a_1, T)] = \int_0^\infty l(a_1, t) f(t) dt \geq \int_0^\infty l(a_2, t) f(t) dt = \mathbb{E}[l(a_2, T)] \qquad (5.2)$$

where f is the density function of T.

The simplest, reasonable, loss function might be specified as follows:

$$l(a_1, t) = c, \forall t > 0$$

$$l(a_2, t) = \left\{ \begin{array}{ll} C & for \ t < \tau \\ -K & for \ t > \tau \end{array} \right. \qquad (5.3)$$

with $K > C > c$. In such a case the optimal decision will be a_2 if and only if

$$\mathbb{E}[l(a_2, T)] = CF(\tau) - K\overline{F}(\tau) \leq c$$

i.e., if and only if

$$\overline{F}(\tau) \geq \frac{C-c}{C+K}. \qquad (5.4)$$

Example 5.3. (The spare parts problem). An apparatus, that is to accomplish a mission of duration τ, contains a component, which is sensitive to a certain type of shock. The subsequent shocks occur according to a Poisson process N_t of intensity μ and each shock causes the failure of the component. We want to determine the optimal number of spare copies of the same component, taking into account that any spare part has a cost c (due to purchase, storage, maintenance, ...), that the gain for the accomplishment of the mission is K, and that the cost caused by the failure of the apparatus previous to τ, is C.

Let M be the maximum number of spares which it is practically possible to purchase.

This is a decision problem where the space of possible actions A is $\{0, 1, 2, ..., M\}$ (choosing $a = n$, means that we provide the apparatus with n spare components).

The unobservable quantity of interest is $W = N_\tau$, the number of shocks before the mission time τ, which coincides with the number of needed spare parts; then $\mathcal{V} = \{0, 1, 2, ...\}$.

The loss function is specified by

$$l(a, w) = c \cdot a + C \cdot \mathbf{1}_{\{w > a\}} - K \cdot \mathbf{1}_{\{w \leq a\}}$$

Since the distribution Π of W is obviously Poisson with mean $\mu\tau$, the expected loss associated with the choice a is

$$\mathbb{E}_\Pi[l(a; W)] = c \cdot a + C \exp\{-\mu\tau\} \sum_{i=a+1}^\infty \frac{(\mu\tau)^i}{i!} - K \exp\{-\mu\tau\} \sum_{i=0}^a \frac{(\mu\tau)^i}{i!}$$

and then

$$\mathbb{E}_{\Pi}\left[l(a;W)\right] - \mathbb{E}_{\Pi}\left[l(a-1;W)\right] = c - (C+K)\exp\{-\mu\tau\}\frac{(\mu\tau)^a}{a!}$$

The optimal action is M if

$$\frac{(\mu\tau)^a}{a!} \geq \frac{c}{C+K}\exp\{\mu\tau\}$$

for $a = 1, 2, ..., M$; otherwise it is the maximum integer a such that

$$\frac{(\mu\tau)^a}{a!} \geq \frac{c}{C+K}\exp\{\mu\tau\}.$$

Example 5.4. (The burn-in problem for a single unit). The burn-in problem for a unit U described in Example 3.69 is a decision problem with

$$A = [0, +\infty], \ \mathcal{V} = [0, \infty)$$

and where we can interpret the choice $a = 0$ as the decision of delivering the unit U to operation immediately and the choice $a = \infty$ as the decision to discard U; the choice $0 < a < \infty$ corresponds to conducting a burn-in of duration a; the unit will be delivered to operations, if it survives the test.

The state of nature in this problem is the overall lifetime (burn-in + operational life) of U, which we denoted by T and whose survival function was denoted by $\overline{F}$. The loss function is

$$l(a, w) = c\mathbf{1}_{\{w<a\}} + C\mathbf{1}_{\{a<w<a+\tau\}} - K\mathbf{1}_{\{w>a+\tau\}} \tag{5.5}$$

More generally we might consider the case where the cost coming from a burn-in of duration a is $c(a)$ and the gain coming from an operative life of duration t yields a gain $K(t)$. It is natural to assume that K is a non-decreasing function, such that $K(t) < 0$ in the neighborhood of the origin and positive for t large enough; the corresponding loss function takes the form

$$l(a, w) = c(a)\mathbf{1}_{\{w<a\}} - \mathbf{1}_{\{w\geq a\}}K(w-a) \tag{5.6}$$

A Bayes optimal burn-in duration then minimizes the expected value

$$\mathbb{E}[l(a;T)] = c(a)F(a) - \int_a^{+\infty} K(w-a)f(w)\,dw. \tag{5.7}$$

The loss function considered in (5.5) is in particular obtained by letting

$$c(a) = c, \forall a > 0$$

$$K(t) = \begin{cases} -C & for\ t < \tau \\ K & for\ t > \tau \end{cases}$$

In such a special case the expected value in (5.7) becomes

$$\mathbb{E}[l(a;T)] = c + (C - c)\overline{F}(a) - (C + K)\overline{F}(a + \tau).$$

The optimal solution for this case will be analyzed in some detail in Section 5.4.

Example 5.5. (Optimal age replacement of a single unit). Here we consider a very simple form of the problem of optimal preventive age replacement of a single unit (see e.g. Barlow and Proschan, 1975; Aven and Jensen, 1999; Gertsbakh, 2000); this allows us to emphasize both the analogies and the differences between such a problem and the problem of optimal burn-in.

U is a unit and its lifetime T has a survival function $\overline{F}$. U is put into operation at time $t = 0$, when it is new. As a consequence of an operative life of length r, we get a gain $K(r)$; however we have a cost Q if the unit fails when still in operation. We can then decide to replace U when it cumulates an age a (if it did not fail before); replacement of U when still in operation causes a cost c $(c < Q)$. We face the problem of optimally choosing the age of replacement a.

We have a decision problem where the space of possible decisions is $\mathcal{V} = [0, \infty]$, $W = T$ and the cost function turns out to be:

$$l(a, w) = -K(w \wedge a) + Q\mathbf{1}_{\{w < a\}} + c\mathbf{1}_{\{w \geq a\}}$$

The expected cost to be minimized then is

$$\mathbb{E}[l(a, T)] = -\mathbb{E}[K(T \wedge a)] + QF(a) + c\overline{F}(a) =$$

$$- \int_0^a K(w)f(w)dw + [c - Q - K(a)]\overline{F}(a) + Q.$$

It is natural to assume that K is a non-decreasing function; if it is also differentiable we obtain that an optimal replacement age a^* must be a solution of the equation

$$r(a^*) = \frac{K'(a^*)}{Q - c}, \tag{5.8}$$

where r denotes the hazard rate function of T.

More precisely, we can conclude as follows:

If $r(0) < \frac{K'(0)}{Q-c}$ and Equation (5.8) is satisfied for some $a^* > 0$ then the smallest such a^* provides the optimal solution; if $r(a) < \frac{K'(a)}{Q-c}$, $\forall a \geq 0$, then we can take $a^* = \infty$, meaning that no preventive replacement is to be planned; if $r(0) \geq \frac{K'(0)}{Q-c}$, then $a^* = 0$ is the optimal choice, meaning that we have to discard the unit immediately. The possibility of a burn-in procedure is to be considered provided $r(a) < \frac{K'(a)}{Q-c}$, for some $a \geq 0$.

Example 5.6. (Inspection procedures for acceptance of lots). Here we consider the classical situation of quality control. A lot of size N of similar units is to be put into assemblies in a production line; the number of defective units in the lot is a random variable W with a distribution Π over the set of possible values $\{0, 1, ..., N\}$.

A simple problem concerns the decision, on the basis of Π, whether to install the lot immediately or to inspect all the units before installation.

In this simple form, the problem is not very realistic. Rather a more common practice consists in inspecting a number n of units and then, on the basis of the observed number of defective units, facing the decision problem described above for the reduced lot of the remaining $N - n$ units.

This shows that a first problem related with this situation is the one of optimally choosing the number n of units to inspect. For details about the decision-theoretic approach to this problem, see e.g. Deming (1982); Barlow and Zhang (1987).

In all the different decision problems of interest in this matter we have that the state of nature W is to be taken as the number of defective units among the non-inspected units.

Example 5.7. (Design of scram systems). A scram device is a unit designed to detect the occurrence of a situation of danger in a given production system and to consequently shut the production system down.

However such a unit can itself fail and behave in an erroneous way; typically it can have two possible failure modes:

· *unsafe:* the unit does not recognize the situation of danger

· *safe:* the unit shuts the production system down when it is not needed.

Due to this, one usually implements a scram system formed of several, say n, units. Since the units are typically similar and all play the same role, we have a situation of symmetry which suggests implementing a $k : n$ system: i.e. the production system will be shut down if and only if at least k out of n units suggest doing so.

Related to the optimal design of a scram system, we can consider different kinds of decision problems; in any case the decision consists in the optimal choice of the numbers n and k (see Clarotti and Spizzichino, 1996).

5.1.1   Statistical decision problems

The problems considered so far are sometimes called "decision problems without observations".

More often however one is interested in a "decision problem with observations" or "statistical decision problem".

The latter means that we have a decision problem as defined above with one more ingredient: the availability of a statistical observation X, before choosing the action a.

X is a random variable and we denote by $\mathcal{X}$ the space of possible values of X.

We consider (W, X) as a random variable with values in the product space $\mathcal{V} \times \mathcal{X}$ and denote by P the joint distribution of the pair (W, X).

We already denoted by Π the (marginal) distribution of W; then, in order to describe the joint distribution of (W, X) we assign the conditional distribution of X, given $\{W = w\}$ (for $w \in \mathcal{V}$); the latter will be denoted by P_w.

Thus a statistical decision problem is specified by the elements $A, \mathcal{V}, l, \mathcal{X}, P$ or, equivalently, by the elements $A, \mathcal{V}, l, \Pi, \mathcal{X}, \{P_w\}_{w \in \mathcal{V}}$; the latter choice is more common and often more convenient in that it is useful, as we shall see, to compare the elements of a decision problem with observations with those of a related decision problem without observations.

It is commonly assumed that the statistical model for the observation X is such that any probability distribution $P_w(w \in \mathcal{V})$ admits a density, which will be denoted by $f(x|w)$ (see also Subsection 5.1.3).

Example 5.8. (Life-testing). We come back to the two-action decision problem of Example 5.2, where T is the lifetime of a unit to be possibly used for the pre-established task.

Now we suppose that n units, with lifetimes $T_1, ..., T_n$, are available and that $T, T_1, ..., T_n$ are exchangeable (but are not stochastically independent). This implies that observing $T_1, ..., T_n$ provides some information about T.

Here, T has the role of the state of nature W, while $X = (T_1, ..., T_n)$. Specializing the form (1.37) of Chapter 1, we have that $f_{\mathbf{X}}(\cdot|t)$ is given by the predictive density

$$f_{\mathbf{T}}(\mathbf{t}|t) = f^{(n)}(\mathbf{t}|t) = \frac{f^{(n+1)}(t, t_1, ..., t_n)}{\int_0^\infty \cdots \int_0^\infty f^{(n+1)}(t, \xi_1, ..., \xi_n)d\xi_1...d\xi_n}.$$

while $f_W(t|T_1 = t_1, ..., T_n = t_n)$ coincides with the predictive density

$$f^{(1)}(t|\mathbf{t}) = \frac{f^{(n+1)}(t, t_1, ..., t_n)}{\int_0^\infty f^{(n+1)}(\xi, t_1, ..., t_n)d\xi}.$$

We can then consider the decision problem with observations whose elements are given as follows:

$$A = \{a_1, a_2\}, \mathcal{V} = [0, \infty), f^{(1)}, \mathcal{X} = [0, \infty)^n, f^{(n)}(\mathbf{t}|t), l$$

where the loss function l is as in (5.3).

After making a life-testing experiment on the n available units, i.e. after observing the statistical result

$$D \equiv \{T_1 = t_1, ..., T_n = t_n\},$$

our state of information on T is described by the conditional density $f_T(t|D) = f^{(1)}(t|\mathbf{t})$.

Conditionally on D, the probability for the unit to survive the mission time τ, is then given by

$$\overline{F}^{(1)}(\tau|t_1, ..., t_n) = \alpha \int_\tau^\infty f^{(n+1)}(\xi, t_1, ..., t_n)d\xi,$$

where for brevity's sake we set α in place of $\frac{1}{\int_0^\infty f^{(n+1)}(\xi, t_1, ..., t_n)d\xi}$.

For any such result D, we can consider the optimal decision corresponding to the new state of information; by adapting formula (5.4), we obtain that a_2 is optimal if and only if the vector $(t_1, ..., t_n)$ is such that

$$\alpha \int_\tau^\infty f^{(n+1)}(\xi, t_1, ..., t_n)d\xi \geq \frac{C-c}{C+K}.$$

We see that the optimal action must be a function of $(t_1, ..., t_n)$ and, under the considered loss function l, it is then

$$a^*(t_1, ..., t_n) = \begin{cases} a_1 & \text{if } \alpha \int_\tau^\infty f^{(n+1)}(\xi, t_1, ..., t_n)d\xi < \frac{C-c}{C+K} \\ a_2 & \text{if } \alpha \int_\tau^\infty f^{(n+1)}(\xi, t_1, ..., t_n)d\xi > \frac{C-c}{C+K} \end{cases} ; \qquad (5.9)$$

we can indifferently take $a^*(t_1, ..., t_n) = a_1$ or $a^*(t_1, ..., t_n) = a_2$ when

$$\alpha \int_\tau^\infty f^{(n+1)}(\xi, t_1, ..., t_n)d\xi = \frac{C-c}{C+K}.$$

Example 5.9. (Acceptance of lots after inspections). Coming back to Example 5.6, we assume that we inspected n units and we found k defective units and $(n-k)$ good units.

After this result, we of course install the good units and discard the k defective units; furthermore we reconsider the problem whether to install the remaining $(N-n)$ units in the lot without further inspection or to inspect all of them.

We see that, at this step, we face a new decision problem without observations, where the essential element of judgment is the conditional distribution $\Pi_{k,n-k}$ that we assess for the number of remaining defective units, given the result observed in the inspection.

Let us now summarize what the above examples say.

In the Bayesian approach it is natural to consider the conditional distribution $\Pi(\cdot|x)$ of the unobservable variable W, given the value x, observed in the statistical experiment performed before taking the decision. Of course $\Pi(\cdot|x)$ can be derived from the initial distribution Π and the statistical model $\{f(x|w)\}_{w \in V}$.

Since $\Pi(\cdot|x)$ completely describes the relevant information we have about W, just after the experiment, we can forget about the observation itself and reconsider a new decision problem, without observations, specified by the elements

$$A, \mathcal{V}, \Pi(\cdot|x), l$$

As a function of the observed data x ($x \in \mathcal{X}$), it is then natural to consider as optimal any action $a^*_{\Pi(\cdot|x)}$, that is a Bayes action against the distribution $\Pi(\cdot|x)$, i.e. such that

$$\mathcal{R}_\Pi(a^*_{\Pi(\cdot|x)}) \leq \mathcal{R}_\Pi(a), \ \forall a \in A. \tag{5.10}$$

We see that, in this way, a decision problem with observations is reduced to an appropriate decision problem without observation.

Remark 5.10. The distinction between decision problems without observations and decision problems with observations gives rise to a possibly useful mental scheme. However, such a distinction is not really substantial in a Bayesian approach. In fact, as we saw, any decision problem with observations reduces to a decision problem without observations after observing the statistical data; on the other hand, any decision problem without observations is such only in that the information contained in past observed data has already been incorporated within the distribution of unobservable quantities of interest.

Notice that $a^*_{\Pi(\cdot|x)}$ is then a mapping from $\mathcal{X}$ to A. Any mapping δ with

$$\delta : \mathcal{X} \to A,$$

in a statistical decision problem, is called *decision function* or *strategy*. A decision function is then a rule associating an action $\delta(x) \in A$ to any possible statistical result $x \in \mathcal{X}$; denote by Δ the set of all possible decision functions.

Fix now a decision function $\delta \in \Delta$ and consider the loss $l(\delta(X); W)$. Notice that, X being a random variable (or vector), $\delta(X)$ is a random action, i.e. $\delta(X)$ takes values in A.

W is also random and then $l(\delta(X); W)$ is a random loss, a scalar random variable; the distribution of $l(\delta(X); W)$ of course depends on the joint distribution of (X, W).

As already mentioned, we usually assume that the conditional distributions of X given $(W = w)$ are fixed and they admit the densities $f(x|w)$.

The marginal (prior) distribution of W is denoted by Π and the marginal (predictive) density of X is given by

$$f_X(x) = \int_{\mathcal{V}} f(x|w) d\Pi(w) \tag{5.11}$$

The expected value of $l(\delta(X); W)$, when it does exist, is expressed by :

$$\mathfrak{r}_\Pi (\delta) \equiv \mathbb{E}_\Pi [l(\delta(X); W)] = \int_{\mathcal{X}} \left[\int_{\mathcal{V}} l(\delta(x); w) f(x|w) d\Pi(w) \right] dx. \qquad (5.12)$$

The quantity $\mathfrak{r}_\Pi (\delta)$ evaluates the risk associated with the strategy δ, against the distribution Π, before getting the information provided by statistical data.

Definition 5.11. $\delta_\Pi^* : \mathcal{X} \to A$ is a *Bayes decision function* (against Π) if the following inequality holds

$$\mathfrak{r}_\Pi(\delta_\Pi^*) \leq \mathfrak{r}_\Pi(\delta)$$

for all $\delta \in \Delta$.

Obviously, when $\mathcal{X}$ and A are not finite, the space Δ is not finite, generally, and then it can happen that no Bayes decision function exists (against some fixed Π).

For a fixed prior distribution Π, consider now the decision problem with observations where the original space $\mathcal{X}$ is replaced by

$$\mathcal{X}_\Pi \equiv \{x \in \mathcal{X} | f_X(x) > 0\}$$

and suppose that a Bayes action $a_{\Pi(\cdot|x)}^*$ exists against the conditional distribution $\Pi(\cdot|x)$, for all $x \in \mathcal{X}_\Pi$. Then we have

Proposition 5.12. *The decision function defined by*

$$\widehat{\delta}_\Pi(x) = a_{\Pi(\cdot|x)}^*, \quad x \in \mathcal{X}_\Pi, \qquad (5.13)$$

is a Bayes strategy against Π.

Proof. For $\delta \in \Delta$, it is

$$\mathfrak{r}_\Pi (\delta) = \int_{\mathcal{X}_\Pi} \left[\int_{\mathcal{V}} l(\delta(x); w) f(x|w) d\Pi(w) \right] dx \geq$$

$$\int_{\mathcal{X}_\Pi} \left[\inf_{a \in A} \int_{\mathcal{V}} l(a; w) f(x|w) d\Pi(w) \right] dx =$$

$$\int_{\mathcal{X}_\Pi} \left[f_X(x) \inf_{a \in A} \int_{\mathcal{V}} l(a; w) \frac{f(x|w)}{f_X(x)} d\Pi(w) \right] dx$$

On the other hand, by taking into account Bayes' formula, we have, $\forall a \in A$,

$$\int_{\mathcal{V}} l(a; w) \frac{f(x|w)}{f_X(x)} d\Pi(w) = \mathbb{E}[l(a; W)|X = x] = \mathcal{R}_{\Pi(\cdot|x)}(a),$$

and then we can write

$$\int_{\mathcal{X}_\Pi} \left[f_X(x) \inf_{a \in A} \int_{\mathcal{V}} l(a; w) \frac{f(x|w)}{f_X(x)} d\Pi(w) \right] dx =$$

$$\int_{\mathcal{X}_\Pi} \left[f_X(x) \int_{\mathcal{V}} l(a^*_{\Pi(\cdot|x)}; w) d\Pi(w|x) \right] dx = \mathfrak{r}_\Pi \left(\widehat{\delta}_\Pi \right).$$

$\square$

Remark 5.13. It is to be noticed that, for $\delta \in \Delta$, the computation of the quantity $\mathfrak{r}_\Pi(\delta)$ involves an integral over the space $\mathcal{X}$ of all the possible statistical results. The evaluation of a decision function δ in terms of the quantity $\mathfrak{r}_\Pi(\delta)$, is then an *a priori* evaluation, which will not take into account the statistical result $x \in \mathcal{X}$ that will be actually observed. Then evaluation of a decision function δ in terms of $\mathfrak{r}_\Pi(\delta)$ should not in general be confused with the *a posteriori* evaluation of actions, conditional on the observation of x.

The *a priori* expected loss, on the other hand, is a useful concept in problems of optimal design of experiments where a decision is really to be taken before getting any statistical observation. For a more detailed discussion and useful examples on these points see Piccinato (1980) and (1993).

Often, as for instance in Example 5.8, we are interested in *predictive statistical decision problems*; this is the case when we have random variables $X_1, ..., X_n, X_{n+1}, ..., X_N$ so that $X \equiv (X_1, ..., X_n)$ is the statistical observation and $W \equiv (X_{n+1}, ..., X_N)$ is considered as the state of nature.

In particular we can be interested in the cases when $(X_1, ..., X_N)$ are exchangeable.

Remark 5.14. Consider in particular the case of a "parametric model" for the random variables $X_1, ..., X_n, X_{n+1}, ..., X_N$ in a predictive statistical decision problem; i.e. the case when $X_1, ..., X_N$ are conditionally i.i.d. given a parameter Θ.

In such a case it may be mathematically convenient to transform the original decision problem into one where Θ appears as the new state of nature. This can be done by defining a new loss function as follows

$$\widehat{l}(a; \theta) = \mathbb{E}[l(a; W) | \Theta = \theta], \theta \in L$$

Remark 5.15. Notice that the operation above is possible any time when the original state of nature W is conditionally independent of the statistical observation X, given a parameter Θ. This explains why the parameter in a statistical model often has also the role of state of nature.

Example 5.16. (Spare parts problem with observations). Let Θ be a non-negative random variable and consider the spare parts problem in the case when the process counting the arrival of shocks is Poisson with intensity θ, conditionally on $\{\Theta = \theta\}$.

Before deciding the number of spare parts, we may want to estimate Θ, and thus we observe the number X_s of shocks which occur in a time interval of length s.

We assume that X_s and N_τ (the number of shocks during the mission time τ) are conditionally independent given $\{\Theta = \theta\}$, and Poisson distributed with means θs and $\theta \tau$, respectively.

We can look at this as a problem where X_s is the observation and Θ is the new state of nature. The corresponding loss function is

$$\widehat{l}(a; \theta) = \mathbb{E}[l(a; W) | \Theta = \theta] =$$

$$c \cdot a - K + (C + K) \exp\{-\theta \tau\} \sum_{i=a+1}^{\infty} \frac{(\theta \tau)^i}{i!} \tag{5.14}$$

Example 5.17. (Life-testing before a burn-in procedure). Suppose that, before facing the decision problem mentioned in Examples 3.69 and 5.4, we obtained a result

$$D = \{T_1 = t_1, ..., T_h = t_h, T_{h+1} > t, ..., T_n > t\}$$

from a life-testing experiment on n units, whose lifetimes are judged to be exchangeable with T.

We can then look at the conditional (predictive) survival function $\overline{F}^{(1)}(t|D)$ of T given D, and consider the optimal solution related to such a distribution. The optimal burn-in duration is a function $a^*(t, n, h; t_1, ..., t_n)$ of the result observed in the life-testing experiment.

Example 5.18. In the problem of designing a scram system, our decision is to be based on the distributions of the times to safe failure and to unsafe failure, for a scram unit, respectively.

Suppose now that the scram units, in our judgement, are similar and not independent; before taking any decision, then, we may want to observe a number, say n, of similar units, each for a total time t. Denoting by U_i and V_i, respectively, the times to safe failure and to unsafe failure of the inspected scram unit i ($i = 1, ..., n$), we observe a special type of *censored data*, in fact what we observe is the set of pairs (T_i, E_i), where

$$T_i = U_i \wedge V_i, \; E_i = \begin{cases} 1 & if \; T_i = U_i < t \\ 0 & if \; T_i = V_i < t \\ \emptyset & if \; T_i > t \end{cases}$$

After observing a result D from this inspection,

$$D = \{(T_1 = t_1, E_1 = e_1), ..., (T_k = t_k, E_k = e_k),$$

$$(T_{k+1} > t, E_{k+1} = \emptyset), ..., (T_n > t, E_n = \emptyset)\},$$

our decisions are to be based on the joint distributions of the times to safe failures and to unsafe failures of scram units, conditional on D.

We have seen that, in the problems of interest in reliability and survival analysis, the following situation is met: for a vector of lifetimes $T_1, ..., T_n, T_{n+1}, ..., T_N$ we have that $X \equiv (T_1, ..., T_n)$ is the statistical observation and $W \equiv (T_{n+1}, ..., T_N)$ is the state of nature.

In other cases W is a vector of residual lifetimes

$$T_{n+1} - r_{n+1}, ..., T_N - r_N.$$

while the observation is of the form

$$D \equiv \{T_1 = t_1, ..., T_h = t_h, T_{h+1} > r_{h+1}, ..., T_N > r_N\}$$

i.e. survival data may appear in the statistical result along with failure data. For this reason we often prefer to denote a decision function by the symbol $\delta(D)$ rather then $\delta(x)$.

5.1.2 Statistical decision problems and sufficiency

In a statistical decision problem, the Bayes strategy, as a function $a^*(D)$ of the observation D, is determined on the basis of the posterior distribution $\Pi_W(\cdot|D)$.

Consider now the case when we can find a function S of the observation which is sufficient for the prediction of W, i.e. such that, for a pair of different set of observations D and $\widehat{D}$, the following implication holds:

$$S(D) = S\left(\widehat{D}\right) \Rightarrow \Pi_W(\cdot|D) = \Pi_W\left(\cdot|\widehat{D}\right).$$

In such a case we obviously have:

$$S(D) = S\left(\widehat{D}\right) \Rightarrow a^*(D) = a^*\left(\widehat{D}\right)$$

This shows the interest in understanding the structure of sufficiency existing in a given decision problem.

Example 5.19. Consider the life testing problem of Example 5.8 in the case when the joint distribution of $(T, T_1, ..., T_n)$ is described by a proportional hazard model.

Let D and $\widehat{D}$ be two different observed data of the form

$$D \equiv \{T_1 = t_1, ..., T_h = t_h, T_{h+1} > r_{h+1}, ..., T_n > r_n\}$$

$$\widehat{D} \equiv \{T_1 = t'_1, ..., T_h = t'_{h'}, T_{h'+1} > r'_{h'+1}, ..., T_n > r'_n\}$$

By the arguments in Example 2.18 we then have

$$a^* (D) = a_2 \Leftrightarrow a^* \left(\widehat{D}\right) = a_2$$

whenever

$$h = h', \sum_{i=1}^{h} R(t_i) = \sum_{i=1}^{h} R(t'_i), \sum_{i=h+1}^{n} R(r_i) = \sum_{i=h+1}^{n} R(r'_i)$$

where $R(\cdot)$ is the cumulative hazard function. In the case $R(t) = t$, we get the well known sufficiency property of the exponential model (see Remark 5.32).

Remark 5.20. The proportional hazard models are the only known models where we can find a fixed-dimension sufficient statistic of data containing survival data for units of different "ages".

In the case when all surviving units in the observed data share the same age, i.e. when, for some $r > 0$, it is

$$D \equiv \{T_1 = t_1, ..., T_h = t_h, T_{h+1} > r, ..., T_N > r\} \tag{5.15}$$

we can have sufficient statistics even for models different from those characterized by proportional hazards; in this respect see the arguments in Costantini and Pasqualucci, (1998).

When the observation is as in (5.15) and the estimation concerns the residual lifetimes

$$T_{h+1} - r, ..., T_N - r.$$

the notion of "dynamic sufficiency" introduced in Section 2.4 can also be useful.

5.1.3   Some technical aspects

For practical reasons, we avoided considering, in this section, a few technical aspects, which are important from a mathematical point of view. Some of them, which might be of interest for non-professional mathematicians, will be mentioned here.

For a given loss function $l(a; w)$ and for a given random variable W (the "state of nature"), we considered the quantities $l(a; W)$ as real random variables ($a \in A$). Furthermore, for a given random variable, or vector, X (the "observation") and for a decision function δ, we considered $\delta(X)$ as a random variable taking values in the space of actions A and $l(\delta(X); W)$ as real random variables.

In order to give the above a rigorous meaning, it is necessary, from a technical point of view, to equip the spaces A, $\mathcal{V}$ and $\mathcal{X}$ with σ-fields of subsets $\mathcal{B}^{(A)}$, $\mathcal{B}^{(\mathcal{V})}$, and $\mathcal{B}^{(\mathcal{X})}$ respectively, and, related to that, we have to require appropriate measurability conditions for the functions $l(\cdot; w)$, $l(a; \cdot)$, δ (see e.g. Billingsley, 1995).

When the space is finite, one tacitly assumes that the σ-field coincides with the family of all its subsets; when the space is a regular region of the set $\mathbb{R}$ of real numbers or of $\mathbb{R}^n$, for some n, the σ-field is the one formed by the Borel sets.

In statistical decision problems (i.e., with statistical observations) the probability distributions $P_\theta (\theta \in L)$ are measures on $(\mathcal{X}, \mathcal{B}^{(\mathcal{X})})$.

As mentioned, it is commonly assumed that any $P_\theta (\theta \in L)$ admits a density, which we denoted by $f(x|\theta)$ (this is not a very precise statement: one should say that $f(x|\theta)$ is a density with respect to a fixed "σ-finite" positive measure μ on $(\mathcal{X}, \mathcal{B}^{(\mathcal{X})})$). Typically, in cases of more frequent interest, however, $\mathcal{X}$ is indeed a regular region of $\mathbb{R}$ or of $\mathbb{R}^n$ and it is tacitly understood that μ is a counting measure or the Lebesgue measure.

The assumption mentioned is, in the statistical literature, referred to by saying that the statistical model $\{P_\theta\}_{\theta \in L}$ is *dominated*.

This condition of domination is really a very important one, since it ensures the following:

If $\pi(\theta)$ is the density of the prior distribution Π with respect to a given σ-finite positive measure μ on $(L, \mathcal{B}^{(L)})$, then also the posterior distribution $\Pi(\theta|X = x)$ admits a density $\pi(\theta|x)$ with respect to μ. The relation between $\pi(\theta|x)$ and $\pi(\theta)$ is described by the familiar Bayes' formula.

When the statistical model is not dominated, we cannot generally use Bayes' formula anymore (at least in the form as we usually know it) to obtain the posterior distribution from the prior distribution and from the knowledge of the statistical model.

In very simple words, we can say that the statistical model is not dominated when, at least for some values x of the statistical observation, the support of the posterior distribution $\Pi(\theta|X = x)$ is different from the one of the prior (i.e., observing x provides some information of "deterministic" type about the parameter).

Example 5.21. Consider the common mode failure model, where we observe

$$T_i = \Theta \wedge \mathcal{E}_i,$$

$\mathcal{E}_1, ..., \mathcal{E}_n$ being i.i.d lifetimes. The family of conditional distributions P_θ of T_i given $(\Theta = \theta)$, $\theta > 0$, is not dominated: given $(\Theta = \theta)$, there is a positive probability that $(T_i = \theta)$, and we cannot derive the conditional distribution of Θ (or of T_j, with $j \neq i$) given $(T_i = t_i)$, by using the Bayes' formula in its common form; even if the prior distribution of Θ admits a density, the posterior distribution of Θ given $(T_i = t_i)$ will not do, since it has a probability mass on the value t_i (see e.g. Macci, 1999, for some discussion on the case of non-domination).

5.2 Stochastic orderings and orderings of decisions

Comparisons between probability distributions can be used to obtain inequalities among Bayes decisions. In fact relevant aspects in the structure of the loss function $l(a_i, w)$ in a decision problem can be combined with properties of stochastic ordering in order to achieve monotonicity properties of decision procedures.

Results of this kind are well known in the classical literature about decision theory (see e.g. Karlin and Rubin, 1956 and Ferguson, 1967).

In this section we shall consider Bayes decisions in the frame of reliability, along with some specific aspects related to the presence of survival data among statistical observations. In particular we concentrate attention on the simple cases when $A = \{a_1, a_2\}$ (two-action problems) and the state of nature W is a scalar quantity ($\mathcal{V} \subseteq \mathbb{R}$).

As a first illustration we start with problems without observations.

Often it is reasonable to assume that the function

$$z(w) = l(a_1, w) - l(a_2, w), \tag{5.16}$$

in a two-action decision problem, is monotone on $\mathcal{V}$ (i.e. one of the two actions becomes more and more preferable than the other, when the value of w becomes bigger and bigger); to fix ideas assume that a_1 and a_2 are indexed in a way so that $z(w)$ is a non-decreasing function.

In this respect we have the following simple result. Let Π_1, Π_2 be two probability distributions for W.

Proposition 5.22. *If $\Pi_1 \leq_{st} \Pi_2$ and $z(w)$ is a non-decreasing function then*

$$a^*_{\Pi_1} = a_2 \Rightarrow a^*_{\Pi_2} = a_2. \tag{5.17}$$

Proof. Let Π be an arbitrary distribution for W. By Equation (5.2), it is $a^*_\Pi = a_2$ if and only if

$$\int_{\mathcal{V}} l(a_1, w) d\Pi(w) - \int_{\mathcal{V}} l(a_2, w) d\Pi(w) = \int_{\mathcal{V}} z(w) d\Pi(w) \geq 0$$

Then we have to show that $\int_{\mathcal{V}} z(w)d\Pi_1(w) \geq 0$ implies $\int_{\mathcal{V}} z(w)d\Pi_2(w) \geq 0$. This implication immediately follows from the assumption that $z(w)$ is a non-decreasing function of w and that $\Pi_1 \leq_{st} \Pi_2$ (recall Theorem 3.12). $\qquad\square$

Example 5.23. In the two-action problem of Example 5.2, we accept the more "risky" decision a_2 (using U for the mission) if and only if the survival function of T, the lifetime of U, is such that $\overline{F}(\tau) \geq \frac{C-c}{C+K}$.

The loss function $l(a_i, t)$ is such that

$$l(a_1, t) - l(a_2, t) = \left\{ \begin{array}{ll} c - C & \text{for } t < \tau \\ c + K & \text{for } t \geq \tau \end{array} \right.$$

is a non-decreasing function. Compare now two distributions $\overline{F}$ and $\overline{G}$ for T such that $\overline{F} \leq_{st} \overline{G}$. It is obvious that if we accept a_2 against $\overline{F}$ then we must accept a_2 against $\overline{G}$, also.

Remark 5.24. In general, in a two-action problem we must implicitly assume that the function $z(w)$ in (5.16) has at least one change of sign (otherwise we would not actually have any significant problem).

Then the condition that $z(w)$ is non-decreasing implies that $z(w)$ changes sign exactly once or, more in detail, that there exist w_0 such that

$$z(w) \leq 0, \text{ for } w \leq w_0; \ z(w) \geq 0, \text{ for } w \geq w_0. \tag{5.18}$$

The latter is obviously a much weaker assumption than monotonicity of $z(w)$.

The following result can be seen as a very special consequence of the sign-variation diminishing property of TP_2 functions; for the reader's use, we provide a direct proof. We notice that in this result only condition (5.18) is actually needed. We consider two prior distributions Π_1 and Π_2 with densities π_1 and π_2, respectively.

Proposition 5.25. *Let the two densities* π_1, π_2 *be positive over* $\mathcal{V}$ *and such that* $\pi_1 \leq_{lr} \pi_2$. *If the condition (5.18) holds, then*

$$a_{\Pi_1}^* = a_2 \Rightarrow a_{\Pi_2}^* = a_2.$$

Proof. The condition $a_{\Pi_i}^* = a_2$ $(i = 1, 2)$ means

$$\int_{-\infty}^{w_0} |z(w)| \pi_i(w) \, dw \leq \int_{w_0}^{\infty} |z(w)| \pi_i(w) \, dw. \tag{5.19}$$

The condition $\pi_1 \leq_{lr} \pi_2$ implies the existence of a value $\overline{w} \in \mathcal{V}$, such that

$$\frac{\pi_2(w)}{\pi_1(w)} \leq 1, \forall w \leq \overline{w} \quad \text{and} \quad \frac{\pi_2(w)}{\pi_1(w)} \geq 1, \forall w \geq \overline{w}. \tag{5.20}$$

In the case $\overline{w} = w_0$, the result is trivial. Let us distinguish the two cases: a) $\overline{w} < w_0$ and b) $\overline{w} > w_0$.

Case a)

For $i = 1, 2$, set

$$R_i \equiv \int_{-\infty}^{\overline{w}} |z(w)|\pi_i(w)\, dw, \quad S_i \equiv \int_{\overline{w}}^{w_0} |z(w)|\pi_i(w)\, dw, \quad T_i \equiv \int_{w_0}^{\infty} |z(w)|\pi_i(w)\, dw,$$

so that the condition $a_{\Pi_i}^* = a_2$ can be rewritten as $R_i + S_i \leq T_i$.

From (5.20), we can write

$$R_2 = \int_{-\infty}^{\overline{w}} |z(w)|\frac{\pi_2(w)}{\pi_1(w)}\pi_1(w)\, dw \leq R_1,$$

$$S_1 = \int_{\overline{w}}^{w_0} |z(w)|\pi_1(w)\, dw \leq \int_{\overline{w}}^{w_0} |z(w)|\frac{\pi_2(w)}{\pi_1(w)}\pi_1(w)\, dw = S_2 \leq$$

$$\frac{\pi_2(w_0)}{\pi_1(w_0)} \int_{\overline{w}}^{w_0} |z(w)|\pi_1(w)\, dw = \frac{\pi_2(w_0)}{\pi_1(w_0)}S_1$$

$$\frac{\pi_2(w_0)}{\pi_1(w_0)}T_1 = \int_{w_0}^{\infty} |z(w)|\pi_1(w)\, dw \leq T_2 = \int_{w_0}^{\infty} |z(w)|\frac{\pi_2(w)}{\pi_1(w)}\pi_1(w)\, dw.$$

We can then conclude that, under the condition $R_1 + S_1 \leq T_1$, we have

$$R_2 + S_2 \leq R_1 + \frac{\pi_2(w_0)}{\pi_1(w_0)}S_1 \leq \frac{\pi_2(w_0)}{\pi_1(w_0)}R_1 + \frac{\pi_2(w_0)}{\pi_1(w_0)}S_1 =$$

$$\frac{\pi_2(w_0)}{\pi_1(w_0)}\left(R_1 + \frac{\pi_2(w_0)}{\pi_1(w_0)}S_1\right) \leq \frac{\pi_2(w_0)}{\pi_1(w_0)}T_1 \leq T_2$$

i.e. the condition $a_{\Pi_2}^* = a_2$.

Case b)

Now we set

$$R_i \equiv \int_{-\infty}^{w_0} |z(w)|\pi_i(w)\, dw, \quad S_i \equiv \int_{w_0}^{\overline{w}} |z(w)|\pi_i(w)\, dw, \quad T_i \equiv \int_{\overline{w}}^{\infty} |z(w)|\pi_i(w)\, dw,$$

so that we have the equivalence

$$a_{\Pi_i}^* = a_2 \Leftrightarrow R_i + S_i \leq T_i$$

In the present case, we can write

$$R_2 = \int_{-\infty}^{w_0} |z(w)| \frac{\pi_2(w)}{\pi_1(w)} \pi_1(w)\, dw \leq \frac{\pi_2(w_0)}{\pi_1(w_0)} R_1$$

$$\frac{\pi_2(w_0)}{\pi_1(w_0)} S_1 \leq S_2 = \int_{w_0}^{\overline{w}} |z(w)| \pi_2(w)\, dw \leq S_1$$

$$T_2 = \int_{\overline{w}}^{\infty} |z(w)| \frac{\pi_2(w)}{\pi_1(w)} \pi_1(w)\, dw \geq T_1.$$

Then, under the condition $R_1 + S_1 \leq T_1$, we obtain

$$R_2 + S_2 \leq \frac{\pi_2(w_0)}{\pi_1(w_0)} R_1 + S_1 \leq R_1 + S_1 \leq T_1 \leq T_2.$$

$\square$

Remark 5.26. Statistical decision problems with two actions correspond, in the standard statistical language, to problems of *testing hypotheses* (see e.g. Lehmann 1986; De Groot, 1970). A two-action problem with a loss function satisfying condition (5.18) can be seen as a problem of testing the hypothesis $\{W \leq w_0\}$ against $\{W > w_0\}$ and Proposition 5.25 can be used in the derivation of the classical result about existence of *uniformly most powerful tests,* for families of densities with monotone likelihood ratio (see Lehmann, 1986).

Here we are essentially interested in extending the analysis of *predictive* two-action life-testing problems, as considered in Example 5.8. More exactly, we consider a two-action decision problem with observations where $A \equiv \{a_1, a_2\}$, T is a non-observable lifetime, and the vector of lifetimes $X = (T_1, ..., T_n)$ is the statistical observation. The loss function is specified by the functions $l(a_i, t)$, so that we incur a random loss $l(a_i, T)$ if we choose the action a_i ($i = 1, 2$).

For two given histories

$$\mathfrak{h}_t = \{T_1 = t_1, ..., T_m = t_m, T_{m+1} > t, ..., T_n > t\}$$

$$\mathfrak{h}'_t = \{T_1 = t'_1, ..., T_{m'} = t'_{m'}, T_{m+1} > t, ..., T_n > t\} \tag{5.21}$$

we want to compare $a^*(\mathfrak{h}_t)$, $a^*(\mathfrak{h}'_t)$, the Bayes decision after observing $\mathfrak{h}_t$ and $\mathfrak{h}'_t$, respectively.

To this purpose, we aim to obtain, for the conditional laws of T given $\mathfrak{h}_t$ and $\mathfrak{h}'_t$, likelihood-ratio comparisons which could be combined with Proposition 5.25. A result in this direction is the following.

Proposition 5.27. *Let $(T_1, ..., T_n, T)$ be MTP_2 and $z(t) = l(a_1, t) - l(a_2, t)$ satisfy the condition (5.18). If $\mathfrak{h}'_t$ is more severe than $\mathfrak{h}_t$, then*

$$a^*(\mathfrak{h}'_t) = a_2 \Rightarrow a^*(\mathfrak{h}_t) = a_2.$$

Proof. First we show that the MTP_2 property for $(T_1, ..., T_n, T)$ implies that $W = (T_1, ..., T_n)$ is increasing in T in the multivariate $\geq_{lr}$ sense.

By the definition 3.20 of multivariate $\geq_{lr}$ comparison, we only must check that, for arbitrary vectors $(t'_1, ..., t'_n)$ and $(t''_1, ..., t''_n)$, it is

$$f_{T_1,...,T_n}(\mathbf{t}' \wedge \mathbf{t}''|t') f_{T_1,...,T_n}(\mathbf{t}' \vee \mathbf{t}''|t'') \geq f_{T_1,...,T_n}(\mathbf{t}'|t') f_{T_1,...,T_n}(\mathbf{t}''|t'') \tag{5.22}$$

for $0 \leq t' \leq t''$.

This is immediate, taking into account that $(T_1, ..., T_n, T)$ is MTP_2 and that (5.22) reduces to

$$f_{T_1,...,T_n,T}(\mathbf{t}' \wedge \mathbf{t}'', t' \wedge t'') f_{T_1,...,T_n,T}(\mathbf{t}' \vee \mathbf{t}'', t' \vee t'') \geq$$

$$f_{T_1,...,T_n,T}(\mathbf{t}', t') f_{T_1,...,T_n,T}(\mathbf{t}'', t'').$$

Now we are in a position to apply Theorem 3.62, where we take $\Theta = T$. This shows that

$$f_T^{(1)}(t|h_t) \geq_{lr} f_T^{(1)}(t|h'_t) \tag{5.23}$$

Then the proof is concluded by applying Proposition 5.25.   □

In Equation (5.21) we considered two observed histories, each containing several survival data of the type $\{T_i > t\}$ for a given t. More generally we can be interested in comparing observed data where surviving individuals have different ages.

In the following we show a result in this direction; we shall restrict attention to the case of infinite extendibility.

Remark 5.28. We recall that, when $T_1, ..., T_n, T$ are conditionally independent given a parameter Θ, sufficient conditions for the MTP_2 property is provided by Theorem 3.59.

Let D and $\widehat{D}$ be two different observations of the type

$$D \equiv \{T_1 = t_1, ..., T_h = t_h, T_{h+1} > r_{h+1}, ..., T_n > r_n\}$$

$$\widehat{D} \equiv \{T_1 = t_1', ..., T_{h'} = t_{h'}', T_{h'+1} > r_{h'+1}', ..., T_n > r_n'\}, \qquad (5.24)$$

say. Similarly to the above, denote by $a^*(D)$ and $a^*(\widehat{D})$ the Bayes decisions after observing D and $\widehat{D}$, respectively, in the predictive two-action problem where T has the role of state of nature, the vector $X \equiv (T_1, ..., T_n)$ has the role of observable variable, and $l(a_i, t)$ $(i = 1, 2)$ is the loss function.

We now consider the case when $T_1, ..., T_n, T$ are conditionally independent and identically distributed given a parameter Θ with conditional density $g(t|\theta)$ and conditional survival function $\overline{G}(t|\theta)$; L denotes the space of possible values of Θ and π_0 the initial density of Θ.

The following result shows that, under conditions analogous to those of Proposition 5.27, we can obtain the same conclusion therein, replacing the histories $\mathfrak{h}_t$ and $\mathfrak{h}_t'$ with observations of the type D and $\widehat{D}$ in (5.24).

Proposition 5.29. *Assume $g(t|\theta)$ to be a TP$_2$ function of t and θ. If $h \leq h'$, $t_i \geq t_i'$ $(i = 1, ..., h)$, $r_i \geq t_i'$ $(i = h+1, ..., h')$, $r_i \geq r_i'$ $(i = h'+1, ..., n)$ and $z(t) = l(a_1, t) - l(a_2, t)$ satisfy the condition (5.18), then*

$$a^*(\widehat{D}) = a_2 \Rightarrow a^*(D) = a_2.$$

Proof. Similarly to what was done in the proof of Proposition 5.27, we must check that $f^{(1)}(t|D)$ and $f^{(1)}(t|\widehat{D})$, the conditional densities of T, given D and $\widehat{D}$, respectively, are such that

$$f^{(1)}(t|D) \geq_{lr} f^{(1)}(t|\widehat{D}). \qquad (5.25)$$

By conditional independence of $T, T_1, ..., T_n$ given Θ, it is

$$f^{(1)}(t|D) = \int_L g(t|\theta)\pi(\theta|D)d\theta, \quad f^{(1)}(t|\widehat{D}) = \int_L g(t|\theta)\pi(\theta|\widehat{D})d\theta.$$

By the TP$_2$ assumption on $g(t|\theta)$, the stochastic comparison (5.25) is implied by the condition

$$\pi(\theta|D) \geq_{lr} \pi(\theta|\widehat{D}) \qquad (5.26)$$

(see Remark 3.16). By Bayes' formula, it is

$$\pi(\theta|D) = g(t_1|\theta)...g(t_h|\theta)\overline{G}(r_{h+1}|\theta)...\overline{G}(r_n|\theta)\pi_0(\theta)$$

$$\pi(\theta|\widehat{D}) = g(t_1'|\theta)...g(t_h'|\theta)g(t_{h+1}'|\theta)...g(t_{h'}'|\theta)\overline{G}(r_{h'+1}'|\theta)...\overline{G}(r_n'|\theta)\pi_0(\theta)$$

and thus the comparison (5.25) is achieved by checking that the product

$$\frac{g(t_1|\theta)...g(t_h|\theta)}{g(t_1'|\theta)...g(t_h'|\theta)} \frac{\overline{G}(r_{h+1}|\theta)...\overline{G}(r_{h'}|\theta)}{g(t_{h+1}'|\theta)...g(t_{h'}'|\theta)} \frac{\overline{G}(r_{h+1}|\theta)...\overline{G}(r_n|\theta)}{\overline{G}(r_{h'+1}'|\theta)...\overline{G}(r_n'|\theta)}$$

in an increasing function of θ. This can, in its turn, be easily obtained by using once again the assumption that $g(t|\theta)$ is TP$_2$ and by taking into account that the latter implies that $\overline{G}(t|\theta)$ is TP$_2$ as well (see Theorems 3.10 and 3.11). $\square$

5.3   Orderings   of   residual   lifetimes   and majorization

In the last section we saw the possible role, in the field of predictive decision problems, of stochastic comparisons between conditional distributions of lifetimes given two different survival and/or failure data D and $\widehat{D}$, for exchangeable lifetimes $T_1, ..., T_n$.

In this section we fix attention on the case when D and $\widehat{D}$ have the same value for the total time on test statistic.

This topic is related to the analysis of some majorization properties for the joint distribution of $T_1, ..., T_n$.

In this respect, it is convenient to start by summarizing some relevant properties of Schur-constant densities. Let us consider then a Schur-constant density

$$f^{(n)}(t_1, ..., t_n) = \phi_n \left(\sum_{i=1}^n t_i \right). \tag{5.27}$$

Let D be an observation of the form

$$D = \{T_1 = t_1, ..., T_h = t_h, T_{h+1} > r_{h+1}, ..., T_n > r_n\} \tag{5.28}$$

and look at the conditional distribution of the residual lifetimes

$$T_{h+1} - r_{h+1}, ..., T_n - r_n,$$

given D.

By combining Equation (5.27) with Equation (2.20) of Chapter 2, we obtain that, for $0 \le r_j \le s_j$, the conditional survival function is

$$P\{T_{h+1} > s_{h+1}, ..., T_n > s_n | D\} =$$

$$\frac{\int_{s_{h+1}}^\infty \cdots \int_{s_n}^\infty \phi_n \left(\sum_{i=1}^n t_i \right) dt_{h+1}...dt_n}{\int_{r_{h+1}}^\infty \cdots \int_{r_n}^\infty \phi_n \left(\sum_{i=1}^n t_i \right) dt_{h+1}...dt_n} \tag{5.29}$$

The latter equation shows the following relevant features of models with Schur-constant densities:

· The conditional distribution of the residual lifetimes $T_{h+1}-r_{h+1},...,T_n-r_n$, given D, has itself a Schur-constant survival function and a Schur-constant density

·· Then, given D, $T_{h+1}-r_{h+1},...,T_n-r_n$ are exchangeable (see Proposition 4.37)

··· Let the observation

$$\widehat{D} = \{T_1 = t_1',...,T_h = t_h', T_{h+1} > r_{h+1}',...,T_n > r_n'\} \tag{5.30}$$

contain the same number of failures, h, as D and have the same value for the total time on test

$$Y' = \sum_{i=1}^{h} t_i' + \sum_{i=h+1}^{n} r_i' = \sum_{i=1}^{h} t_i + \sum_{i=h+1}^{n} r_i = Y; \tag{5.31}$$

then the conditional distributions of the residual lifetimes of the surviving individuals, given D and $\widehat{D}$ respectively, do coincide.

We can thus conclude with the following

Proposition 5.30. *Let $f^{(n)}$ be Schur-constant. Then the conditional distribution of the residual lifetimes, given D, is exchangeable and the pair (h,Y) is a sufficient statistic.*

Remark 5.31. Exchangeability of the residual lifetimes given observations D as in (5.28) is not only necessary, but also a sufficient condition for the joint density $f^{(n)}$ to be Schur-constant (see again Proposition 4.37).

Remark 5.32. For the case when $T_1,...,T_n$ are conditionally i.i.d. exponential, the sufficiency property in Proposition 5.30 is a special case of that shown in Example 5.19.

We want now turn to the main problem of this section and then consider the case of a density which is not necessarily Schur-constant.

Let D and $\widehat{D}$ be two different observations as in (5.28) and (5.30), respectively.

We naturally expect that, even if D and $\widehat{D}$ satisfy the equality (5.31), i.e. if they contain the same number of failures and have the same value for the total time on test statistic, the conditional distributions, given D and $\widehat{D}$, of vectors of the residual lifetimes $(T_{h+1} - r_{h+1},...,T_n - r_n)$ and $(T_{h+1} - r_{h+1}',...,T_n - r_n')$ are different.

Furthermore, such conditional distributions are not exchangeable (if $f^{(n)}$ is in particular Schur-concave or Schur-convex, the marginal distributions are ordered as explained in Proposition 4.33).

Next, we analyze conditions, for the joint distribution of the lifetimes, under which the two distributions can be compared in some stochastic sense, at least for some pairs D and $\widehat{D}$ satisfying suitable conditions of majorization.

For simplicity's sake we compare the one-dimensional conditional distributions, given D and $\widehat{D}$, respectively, of $T_j - r_j$ and $T_j - r'_j$ for $j = h+1, ..., n$; due to exchangeability there is no loss of generality in fixing $j = n$.

We start analyzing the case of observations of only survivals, i.e. the case $h = 0$, furthermore we take $r_n = r'_n = r$, for some $r \geq 0$.

Let us then look at the conditional survival functions $\overline{F}_{T_n - r}(\cdot | D)$ and $\overline{F}_{T_n - r}(\cdot | \widehat{D})$ where, for two sets of ages $(r_1, ..., r_{n-1}, r)$ and $(r'_1, ..., r'_{n-1}, r)$, it is

$$D \equiv \{T_1 > r_1, ..., T_{n-1} > r_{n-1}, T_n > r\}, \quad \widehat{D} \equiv \{T_1 > r'_1, ..., T_{n-1} > r'_{n-1}, T_n > r\}$$

Remember that we generally want D and $\widehat{D}$ to satisfy the equality (5.31) which here reduces to

$$\sum_{i=1}^{n-1} r_i = \sum_{i=1}^{n-1} r'_i;$$

in particular we assume, say, the condition

$$\left(r'_1, ..., r'_{n-1} \right) \preceq \left(r_1, ..., r_{n-1} \right). \tag{5.32}$$

For this case we want to compare $\overline{F}_{T_n - r}(\cdot | D)$ and $\overline{F}_{T_n - r}(\cdot | \widehat{D})$ in the sense of the (one-dimensional) hazard rate ordering. It is not restrictive in practice to limit attention to the case $r = 0$, so that

$$D \equiv \{T_1 > r_1, ..., T_{n-1} > r_{n-1}\}, \quad \widehat{D} \equiv \{T_1 > r'_1, ..., T_{n-1} > r'_{n-1}\}$$

In view of Theorem 3.10, we look for conditions such that the ratio

$$\frac{\overline{F}_{T_n}(t | D)}{\overline{F}_{T_n}(t | \widehat{D})}$$

is a monotone function of t, or in particular an increasing function, say. Since

$$\overline{F}_{T_n}(t | D) = \frac{\overline{F}^{(n)}(r_1, ..., r_{r-1}, t)}{\overline{F}^{(n-1)}(r_1, ..., r_{r-1})},$$

and a similar expression is valid for $\overline{F}_{T_n}(t | \widehat{D})$, we can equivalently consider the condition

$$\frac{\overline{F}^{(n)}(r_1, ..., r_{r-1}, t)}{\overline{F}^{(n)}(r'_1, ..., r'_{r-1}, t)} \quad \text{increasing function of } t. \tag{5.33}$$

When $\overline{F}^{(n)}$ admits the first-order partial derivatives with respect to its variables, the latter condition can be written as

$$\frac{\partial}{\partial t}\overline{F}^{(n)}(r_1,...,r_{r-1},t) \geq \frac{\partial}{\partial t}\overline{F}^{(n)}(r'_1,...,r'_{r-1},t)\frac{\overline{F}^{(n)}(r_1,...,r_{r-1},t)}{\overline{F}^{(n)}(r'_1,...,r'_{r-1},t)}.$$

or as

$$\left|\frac{\partial}{\partial t}\overline{F}^{(n)}(r_1,...,r_{r-1},t)\right| \leq \left|\frac{\partial}{\partial t}\overline{F}^{(n)}(r'_1,...,r'_{r-1},t)\right|\frac{\overline{F}^{(n)}(r_1,...,r_{r-1},t)}{\overline{F}^{(n)}(r'_1,...,r'_{r-1},t)}, \quad (5.34)$$

since $\overline{F}^{(n)}$ is a non-increasing function of its coordinates and then

$$\frac{\partial}{\partial t}\overline{F}^{(n)}(r_1,...,r_{r-1},t) \leq 0.$$

Remark 5.33. Assume the existence of a continuous joint density $f^{(n)}$ for $\overline{F}^{(n)}$. Then it is

$$\left|\frac{\partial}{\partial t}\overline{F}^{(n)}(r_1,...,r_{n-1},t)\right| = \int_{r_1}^{\infty} ... \int_{r_{n-1}}^{\infty} f^{(n)}(t_1,...,t_{n-1},t)dt_1...dt_{n-1}.$$

By applying Theorem 4.30, we see that Schur-concavity of $f^{(n)}$ is a sufficient condition for Schur-concavity of both $\overline{F}^{(n)}$ and $|\frac{\partial}{\partial t}\overline{F}^{(n)}(r_1,...,r_{n-1},t)|$.

On the other hand, when $\overline{F}^{(n)}(r_1,...,r_{r-1},t)$ is Schur-concave, the inequality (5.34) is a stronger condition than Schur-concavity of $|\frac{\partial}{\partial t}\overline{F}^{(n)}(r_1,...,r_{r-1},t)|$.

The following example presents the case of survival functions $\overline{F}^{(n)}$, which satisfy the condition (5.33), without being differentiable nor being necessarily Schur-concave.

Example 5.34. (Common mode failure models). Extending Examples 4.26 and 2.22, consider joint survival functions of the form

$$\overline{F}^{(n)}(s_1,...,s_n) = \overline{G}(s_1) \cdot ... \cdot \overline{G}(s_n) \cdot \overline{H}\left(\max_{1\leq i\leq n} s_i\right)$$

where $\overline{G}$ and $\overline{H}$ are one-dimensional survival functions. The condition (5.33) becomes

$$\frac{\overline{H}(z \vee v)}{\overline{H}(z' \vee v)} \text{ increasing in } v$$

where we set $z = \max_{1\leq i\leq n-1} r_i$, $z' = \max_{1\leq i\leq n-1} r'_i$. It is easy to see that this happens, irrespective of the choice of $\overline{G}$ and $\overline{H}$ (see Exercise 5.67). Notice that log-concavity of $\overline{G}$ implies Schur-concavity of $\overline{F}^{(n)}$.

Now we consider the case when $T_1, T_2, ..., T_n$ are conditionally independent identically distributed, given a scalar parameter Θ, i.e. when the joint survival function is of the form

$$\overline{F}^{(n)}(s_1, ..., s_{n-1}, s_n) = \int_L \prod_{j=1}^n \overline{G}(s_j|\theta)\pi_0(\theta)d\theta, \qquad (5.35)$$

with $L \subset \mathbb{R}$. For this case we have the following easy result.

Proposition 5.35. *Let $\overline{G}(s|\theta)$ admit a partial derivative w.r.t. θ and satisfy the conditions*
i) $\frac{\overline{G}(s|\theta)}{\overline{G}(s'|\theta)}$ increasing (decreasing) as a function of s and θ, for $s < s'$
ii) $\frac{\partial}{\partial\theta}\log\overline{G}(s|\theta)$ concave (convex) as a function of s, $\forall\theta \in L$.
Then (5.33) holds.

Proof. We can limit to analyze the case $\frac{\overline{G}(s|\theta)}{\overline{G}(s'|\theta)}$ increasing, $\frac{\partial}{\partial\theta}\log\overline{G}(s|\theta)$ concave. Due to conditional independence given Θ, we have

$$\overline{F}_{T_n}(t|D) = \int_L \overline{G}(t|\theta)\pi(\theta|D)d\theta, \quad \overline{F}_{T_n}(t|\widehat{D}) = \int_L \overline{G}(t|\theta)\pi(\theta|\widehat{D})d\theta$$

where

$$\pi(\theta|D) \propto \pi_0(\theta)\prod_{j=1}^{n-1}\overline{G}(r_j|\theta), \quad \pi(\theta|\widehat{D}) \propto \pi_0(\theta)\prod_{j=1}^{n-1}\overline{G}(r_j'|\theta)$$

Resorting to Remark 3.16, we see that, if $\frac{\overline{G}(s|\theta)}{\overline{G}(s'|\theta)}$ is increasing, we have that $\frac{\overline{F}_{T_n}(t|D)}{\overline{F}_{T_n}(t|\widehat{D})}$ is increasing in t when the ratio $\frac{\pi(\theta|D)}{\pi(\theta|\widehat{D})}$ is a decreasing function of θ, i.e. when $\frac{\prod_{j=1}^{n-1}\overline{G}(r_j|\theta)}{\prod_{j=1}^{n-1}\overline{G}(r_j'|\theta)}$ is a decreasing function of θ.
Now notice that

$$\frac{\partial}{\partial\theta}\frac{\prod_{j=1}^{n-1}\overline{G}(r_j|\theta)}{\prod_{j=1}^{n-1}\overline{G}(r_j'|\theta)} =$$

$$\left[\prod_{j=1}^{n-1}\overline{G}(r_j|\theta)\right]\left[\prod_{j=1}^{n-1}\overline{G}(r_j'|\theta)\right]\left[\sum_{j=1}^{n-1}\frac{\frac{\partial}{\partial\theta}\overline{G}(r_j|\theta)}{\overline{G}(r_j|\theta)} - \sum_{j=1}^{n-1}\frac{\frac{\partial}{\partial\theta}\overline{G}(r_j'|\theta)}{\overline{G}(r_j'|\theta)}\right].$$

Then (recall Proposition 4.12) the condition ii) implies

$$\frac{\partial}{\partial\theta}\frac{\prod_{j=1}^{n-1}\overline{G}(r_j|\theta)}{\prod_{j=1}^{n-1}\overline{G}(r_j'|\theta)} \leq 0.$$

$\square$

Example 5.36. Continuing Examples 2.18 and 4.22, let us consider the case of proportional hazard models, where (5.35) holds with $L = [0, \infty)$ and

$$\overline{G}(s|\theta) = \exp\{-\theta R(s)\},$$

$R(s)$ being an increasing function. It is

$$\frac{\overline{G}(s|\theta)}{\overline{G}(s'|\theta)} = \exp\{\theta[R(s') - R(s)]\}$$

and then $\frac{\overline{G}(s|\theta)}{\overline{G}(s'|\theta)}$ is increasing, for $s < s'$; furthermore

$$\frac{\partial}{\partial \theta} \log \overline{G}(s|\theta) = -R(s)$$

and then we obtain that $\frac{\prod_{j=1}^{n-1} \overline{G}(r_j|\theta)}{\prod_{j=1}^{n-1} \overline{G}(r_j'|\theta)}$ is decreasing (or increasing) when $R(s)$ is convex (or concave), which corresponds to $\overline{F}^{(n)}$ being Schur-concave (or Schur-convex).

Remark 5.37. When the function $R(\cdot)$ in the example above admits a derivative $\rho(\cdot)$ and this is non-decreasing, $\overline{F}^{(n)}$ is Schur-concave and differentiable and the inequality (5.34) holds. However, in such a case, something even stronger happens. Consider in fact the conditional densities

$$f_{T_n}(t|D) = \int_0^\infty g(t|\theta)\pi(\theta|D)d\theta, \; f_{T_n}(t|\widehat{D}) = \int_0^\infty g(t|\theta)\pi(\theta|\widehat{D})d\theta$$

where

$$g(t|\theta) = \theta\rho(t)\exp\{-\theta R(t)]\}$$

and notice that $R(t)$ being convex (and then $\overline{F}^{(n)}$ Schur-concave) also means that the ratio

$$\frac{g(t|\theta)}{g(t'|\theta)} = \frac{\rho(t)}{\rho(t')} \exp\{\theta[R(t') - R(t)]\}$$

is increasing in θ for $t < t'$.

By taking into account the corresponding monotonicity property for $\frac{\pi(\theta|D)}{\pi(\theta|\widehat{D})}$, we can then conclude that even the stronger comparison

$$f_{T_n}(t|D) \geq_{lr} f_{T_n}(t|\widehat{D}) \tag{5.36}$$

holds.

5.3.1   The case of observations containing failure data

Now we analyze the case when the observations D and $\widehat{D}$ also contain failure data for the lifetimes $T_1, ..., T_{n-1}$; in this case we need to develop arguments which directly involve the conditional densities and we compare the two conditional distributions in the sense of (one-dimensional) likelihood-ratio ordering.

Consider two observations

$$D = \{T_1 = t_1, ..., T_h = t_h, T_{h+1} > r_{h+1}, ..., T_{n-1} > r_{n-1}\},$$

$$\widehat{D} = \{T_1 = t_1', ..., T_h = t_h', T_{h+1} > r_{h+1}', ..., T_{n-1} > r_{n-1}'\} \tag{5.37}$$

such that

$$\sum_{i=1}^{h} t_i' = \sum_{i=1}^{h} t_i, \quad \sum_{i=h+1}^{n-1} r_i' = \sum_{i=h+1}^{n-1} r_i. \tag{5.38}$$

and look at the condition

$$f_{T_n}(\cdot|D) \geq_{lr} f_{T_n}(\cdot|\widehat{D}). \tag{5.39}$$

First, we focus again attention on the case of conditionally i.i.d. lifetimes, where the joint density is of the form

$$f^{(n)}(t_1, ..., t_{n-1}, t_n) = \int_L \prod_{j=1}^{n} g(t_j|\theta)\pi_0(\theta)d\theta. \tag{5.40}$$

and then

$$f_{T_n}(t|D) = \int_L g(t|\theta)\pi(\theta|D)d\theta, \quad f_{T_n}(\cdot|\widehat{D}) = \int_L g(t|\theta)\pi(\theta|\widehat{D})d\theta \tag{5.41}$$

with

$$\pi(\theta|D) \propto \pi_0(\theta) \prod_{j=1}^{h-1} g(t_j|\theta) \prod_{j=h+1}^{n-1} \overline{G}(r_j|\theta),$$

$$\pi(\theta|\widehat{D}) \propto \pi_0(\theta) \prod_{j=1}^{h-1} g(t_j'|\theta) \prod_{j=h+1}^{n-1} \overline{G}(r_j'|\theta).$$

As usual we set $\overline{G}(s|\theta) = \int_s^\infty g(t)dt$ and assume $L \subset \mathbb{R}$.

Similarly to Proposition 5.35, this time we have the following result, for which we assume the existence of partial derivatives of both $g(t|\theta)$ and $\overline{G}(s|\theta)$ with respect to the variable θ.

Proposition 5.38. *Let* $\frac{g(t|\theta)}{g(t'|\theta)}$ *be increasing in* θ, *for* $t < t'$ *and let* $\frac{\partial}{\partial\theta}\log g(t|\theta)$, $\frac{\partial}{\partial\theta}\log \overline{G}(s|\theta)$ *be concave functions of* t *and* s, *respectively. Then*

$$f_{T_n}(\cdot|D) \geq_{lr} f_{T_n}(\cdot|\widehat{D})$$

whenever $\mathbf{t}' \preceq \mathbf{t}$, $\mathbf{r}' \preceq \mathbf{r}$.

Proof. By taking into account the identities in (5.41) and reasoning as in the proof of Proposition 5.35, the proof reduces to showing that $\frac{\pi(\theta|D)}{\pi(\theta|\widehat{D})}$ is an increasing function of θ. We now have

$$\frac{\pi(\theta|D)}{\pi(\theta|\widehat{D})} \propto \frac{\prod_{j=1}^{h-1} g(t_j|\theta) \prod_{j=h+1}^{n-1} \overline{G}(r_j|\theta)}{\prod_{j=1}^{h-1} g(t'_j|\theta) \prod_{j=h+1}^{n-1} \overline{G}(r'_j|\theta)}.$$

Similarly to what was done in the proof of Proposition 5.35, the ratio $\frac{\pi(\theta|D)}{\pi(\theta|\widehat{D})}$ is then easily seen to be increasing by developing the partial derivatives of $\frac{\prod_{j=1}^{h} g(t_j|\theta)}{\prod_{j=1}^{h} g(t'_j|\theta)}$, $\frac{\prod_{j=h+1}^{n-1} \overline{G}(r_j|\theta)}{\prod_{j=h+1}^{n-1} \overline{G}(r'_j|\theta)}$ with respect to θ and by taking into account the assumptions of concavity for $\frac{\partial}{\partial\theta}\log g(t|\theta)$ and $\frac{\partial}{\partial\theta}\log\overline{G}(s|\theta)$. $\square$

Example 5.39. As in Remark 5.37, take $L = [0,\infty)$ and $g(t|\theta)$ of the form

$$g(t|\theta) = \theta\rho(t)\exp\{-\theta R(t)\}.$$

It is easy to see that the conditions in Proposition 5.38 hold when $\rho(t)$ is increasing.

For this special kind of model where $T_1, ..., T_n$ are conditionally independent with a joint density of the form

$$f^{(n)}(t_1,...,t_n) = \int_0^\infty \theta^n \rho(t_1)...\rho(t_n)\exp\{-\theta\sum_{i=1}^n R(t_i)\}\pi_0(\theta)\,d\theta. \qquad (5.42)$$

(proportional hazard models with differentiable cumulative hazard rate functions R), it is interesting to notice that the following conditions are all equivalent:

i) $\rho(t)$ is increasing,

ii) $\overline{F}(\cdot)$ is Schur-concave

iii) $\frac{\partial}{\partial\theta}\log\overline{G}(s|\theta) = -R(t)$ is concave,

iv) $\frac{\partial}{\partial\theta}\log g(t|\theta) = \frac{1}{\theta} - R(t)$ is concave.

At the cost of some more assumptions and technicalities the above arguments can be extended to the case when $T_1, ..., T_n$ are conditionally i.i.d. given a multidimensional parameter Θ.

Now we turn however to consider more general cases of exchangeable densities $f^{(n)}$ and we limit to observations exclusively containing failure data:

$$D = \{T_1 = t_1, ..., T_{n-1} = t_{n-1}\}, \widehat{D} = \{T_1 = t'_1, ..., T_{n-1} = t'_{n-1}\}. \qquad (5.43)$$

Notice that the condition (5.39) becomes

$$\frac{f_{T_n}(t|T_1 = t_1, ..., T_{n-1} = t_{n-1})}{f_{T_n}(t|T_1 = t'_1, ..., T_{n-1} = t'_{n-1})} \text{ increasing function of } t,$$

which is equivalent to the condition

$$\frac{f^{(n)}(t_1, ..., t_{n-1}, t)}{f^{(n)}(t'_1, ..., t'_{n-1}, t)} \text{ increasing function of } t. \qquad (5.44)$$

When $f^{(n)}$ admits the first-order partial derivative $\frac{\partial}{\partial t} f^{(n)}(t_1, ..., t_{n-1}, t)$, the latter is in turn equivalent to

$$\frac{\partial}{\partial t} f^{(n)}(t_1, ..., t_{n-1}, t) f^{(n)}(t'_1, ..., t'_{n-1}, t) \geq \frac{\partial}{\partial t} f^{(n)}(t'_1, ..., t'_{n-1}, t) f^{(n)}(t_1, ..., t_{n-1}, t). \qquad (5.45)$$

Consider now

$$\mathbf{t}' \equiv (t'_1, ..., t'_{n-1}) \preceq \mathbf{t} \equiv (t_1, ..., t_{n-1}).$$

We notice that the inequality (5.45), when $\frac{\partial}{\partial t} f^{(n)}(t'_1, ..., t'_{n-1}, t) \leq 0$, is trivially equivalent to

$$\left| \frac{\partial}{\partial t} f^{(n)}(t_1, ..., t_{r-1}, t) \right| \leq \left| \frac{\partial}{\partial t} f^{(n)}(t'_1, ..., t'_{r-1}, t) \right| \frac{f^{(n)}(t_1, ..., t_{r-1}, t)}{f^{(n)}(t'_1, ..., t'_{r-1}, t)}$$

and consider the class of the real-valued functions $W(t_1, ..., t_{n-1}, t)$ satisfying the property

$$
\begin{aligned}
|W(t_1, ..., t_{n-1}, t)| &\leq |W(t'_1, ..., t'_{n-1}, t)| \frac{f^{(n)}(t_1, ..., t_{n-1}, t)}{f^{(n)}(t'_1, ..., t'_{n-1}, t)} && \text{for } W(t'_1, ..., t'_{n-1}, t) < 0 \\
W(t_1, ..., t_{r-1}, t) &\geq W(t'_1, ..., t'_{r-1}, t) && \text{for } W(t'_1, ..., t'_{n-1}, t) \geq 0
\end{aligned}
$$

$$(5.46)$$

for $(t'_1, ..., t'_{n-1}) \preceq (t_1, ..., t_{n-1})$.
The following result can easily be proved (see Exercise 5.68).

Proposition 5.40. *Let $f^{(n)}$ be Schur-concave and $\frac{\partial}{\partial t} f^{(n)}(t_1, ..., t_{n-1}, t)$ satisfy the conditions in (5.46). Then, $(t'_1, ..., t'_{n-1}) \preceq (t_1, ..., t_{n-1})$ implies*

$$f_{T_n}(\cdot | D) \geq_{lr} f_{T_n}(\cdot | \widehat{D}).$$

Remark 5.41. The hypotheses in Proposition 5.40 are rather strong; however, they are sufficient but not necessary for the result; in some cases when the joint density is Schur-concave, the condition (5.44) can be substantially obtained without conditions of differentiability.

A case of interest is shown in the next example.

Example 5.42. (Linear breakdown model). Consider the joint density

$$f^{(n)}(t_1, ..., t_n) = \frac{\theta^n}{n!} \exp\{-\theta \cdot t_{(n)}\}$$

(see Example 2.50). In this case it is

$$\frac{f^{(n)}(t_1, ..., t_{n-1}, u)}{f^{(n)}(t'_1, ..., t'_{n-1}, u)} = \exp\{\theta \left[u \vee z' - u \vee z \right]\}$$

where we let $z \equiv \max(t_1, ..., t_{n-1})$, $z' \equiv \max(t'_1, ..., t'_{n-1})$.

Since $\max(t_1, ..., t_{n-1})$ is a Schur-convex function, we have that, as already noticed, $f^{(n)}$ is Schur-concave and that, for $(t'_1, ..., t'_{n-1}) \preceq (t_1, ..., t_{n-1})$, it is $z' \leq z$; whence the desired monotonicity property of $\frac{f^{(n)}(t_1, ..., t_{n-1}, u)}{f^{(n)}(t'_1, ..., t'_{n-1}, u)}$ readily follows.

Remark 5.43. Arguments above can be reformulated to deal with some other cases by reversing the directions of inequalities and by suitably replacing the condition of Schur-concavity with that of Schur-convexity and vice versa.

A natural application of arguments here is, for instance, the life-testing problem with two actions. For the case when the units are conditionally i.i.d. exponential, the following is a well known (and very useful) fact: sometimes we prefer to test a few units for a long time, other times we prefer, for various cost reasons, to test a large number of units, each for a short time.

When the total number of failures and the total time on test are the same in the two different experiments, we reach the same information concerning the residual lifetimes (of both surviving and new units). This means that we must do the same prediction on the behavior of those units and then any choice between two different actions concerning them must also be necessarily the same (see e.g. Barlow and Proschan, 1988).

We can then conclude that, as an important issue of conditional exponentiality of similar units, two different observations lead to the same Bayes decision, provided the total time on test and the number of failures is the same.

The same argument can also be repeated for any model with a Schur-constant density, slightly more generally.

It is also important to stress that, in Schur-constant models, the conditional distribution for a residual lifetime, given a set of failure and survival data, does not depend on the age of the surviving unit of interest (where age 0 in particular means that the unit is new). This implies that, in predictive decision problems, we do not have to distinguish between new units and used units (provided they have not failed yet).

When we drop the assumption that the joint density is Schur-constant, we expect that two different observations contain different information, even though they present the same number of failures and the same total time on test statistics. It is then natural to wonder which of the two observations is more favorable, to one of the two actions (say a_1), than the other one.

The problem here is in some sense analogous, even though different, from what we considered in the previous section.

After directing the reader's attention to the uses of notions of stochastic comparisons in decision problems with observations, therein we ordered the conditional distributions, given two sets of data comparable in the sense of less severeness.

Here we can repeat arguments analogous to those therein. The difference is in that we compare two sets of failure or survival data, which present the same value for the total time on test statistic and then which cannot be comparable in the sense of less severeness.

Example 5.44. Let $T_1, T_2, ..., T_n$ be lifetimes with joint density $f^{(n)}$ and consider the lifetesting problem for T_n, with loss function specified in (5.3). $X = (T_1, ..., T_{n-1})$ is the statistical observation and we denote by $\mathcal{X}_{a_2}$ the region of acceptance of the more risky action a_2, for the Bayes decision function, i.e. we set

$$\mathcal{X}_{a_2} = \{(t_1, ..., t_{n-1}) \, | a^* (t_1, , ..., t_{n-1}) = a_2\}.$$

It is then (see also (5.9))

$$\mathcal{X}_{a_2} = \{(t_1, ..., t_{n-1}) \, | \int_\tau^\infty f_{T_n} (t|t_1, ..., t_{n-1}) \geq \frac{C - c}{C + K}\}.$$

If $f^{(n)}$ is such that the condition (5.44) holds for $(t'_1, ..., t'_{n-1}) \preceq (t_1, ..., t_{n-1})$, then we have that $\mathcal{X}_{a_2}$ is a *Schur-convex* set, i.e. its indicator function is *Schur-convex*.

Remark 5.45. Let X be a lifetime with a density f and Y a lifetime with a density g and consider, for $r > 0$, the conditional densities

$$f_{X-r} (t|X > r) = \frac{f(t + r)}{\int_r^\infty f(x)dx}, \quad g_{Y-r} (t|Y > r) = \frac{g(t + r)}{\int_r^\infty g(x)dx}.$$

Then it is immediately seen that $f \leq_{lr} g$ implies $f_{X-r}(\cdot|X > r) \leq_{lr} g_{Y-r}(\cdot|Y > r)$. This shows that, under the assumptions of Proposition 5.40, we also have, for any $r > 0$, the implication

$$(t_1', ..., t_{n-1}') \preceq (t_1, ..., t_{n-1}) \Rightarrow \overline{F}_{T_n - r}(\cdot|\widehat{D}) \leq_{lr} \overline{F}_{T_n - r}(\cdot|D), \qquad (5.47)$$

where now

$$D = \{T_1 = t_1, ..., T_{n-1} = t_{n-1}, T_n > r\},$$

$$\widehat{D} = \{T_1 = t_1', ..., T_{n-1} = t_{n-1}', T_n > r\}.$$

Example 5.46. Some ideas, intrinsic to the arguments of this section, can be well illustrated by a further analysis of proportional hazard models.

Take $T_1, ..., T_n$ conditionally independent with a joint density of the form (5.42)

We assume $\rho(t) = R'(t)$ to be monotone; recall that ρ being increasing means that the joint survival function is Schur-concave and that the stochastic comparison

$$\overline{F}_{T_n - r}(\cdot|\widehat{D}) \leq_{lr} \overline{F}_{T_n - r}(\cdot|D)$$

holds, for the residual lifetime $T_n - r$, with

$$D = \{T_1 = t_1, ..., T_h = t_h, T_{h+1} > r_{h+1}..., T_{n-1} > r_{n-1}, T_n > r\},$$

$$\widehat{D} = \{T_1 = t_1', ..., T_h = t_h', T_{h+1} > r_{h+1}', ..., T_{n-1} > r_{n-1}'T_n > r\}$$

$$\mathbf{t}' \prec \mathbf{t}, \mathbf{r}' \prec \mathbf{r}.$$

If ρ is decreasing the comparison

$$\overline{F}_{T_n - r}(\cdot|D) \leq_{lr} \overline{F}_{T_n - r}(\cdot|\widehat{D})$$

on the contrary holds. Take now

$$r_{n-1}' = r_{n-1}$$

and suppose $r_{n-1} < r$, say, and also compare the conditional survival probabilities

$$P\{T_{n-1} - r_{n-1} > \tau|D\}, P\{T_n - r > \tau|D\},$$

$$P\{T_{n-1} - r_{n-1} > \tau|\widehat{D}\}, P\{T_n - r > \tau|\widehat{D}\}.$$

By Proposition 4.33, it is

$$P\{T_{n-1} - r_{n-1} > \tau|D\} \geq P\{T_n - r > \tau|D\}$$

or

$$P\{T_{n-1} - r_{n-1} > \tau|D\} \leq P\{T_n - r > \tau|D\}$$

according to ρ being increasing or decreasing. The same happens for conditional probabilities given $\widehat{D}$.

This then means that if, between two surviving units, we prefer the less aged one, we should also evaluate the observation D "more optimistic" than $\widehat{D}$.

5.4   Burn-in problems for exchangeable lifetimes

Here we consider the problem of optimally choosing a burn-in procedure for n units $U_1, ..., U_n$ with exchangeable lifetimes $T_1, ..., T_n$ and shall also see some application of the likelihood ratio orderings.

Throughout, the assumptions described below are made (see also Spizzichino, 1991, Runggaldier, 1993b).

We assume that the burn-in starts simultaneously for all the units at time 0 when they are new (of age 0); at any time $t > 0$, all the working units share the same age t.

When the burn-in is terminated, all the surviving units are delivered to operation.

Furthermore it is assumed that the units undergo the same stress level, both during burn-in and when in operation.

The history of what happened up to any time t, before the burn-in experiment is terminated, is a dynamic history of the form:

$$\mathfrak{h}_t = \{H_t = h, T_{(1)} = t_1, ..., T_{(h)} = t_h, T_{(h+1)} > t\} \tag{5.48}$$

as in (2.40); $T_{(1)}, ..., T_{(n)}$ are the order statistics of $T_1, ..., T_n$ and H_t is the process counting the number of failures observed up to t (recall position (2.57)). Here we emphasize the role of the variable H_t and add the index t, to emphasize dependence on time.

We assume that we have complete information about what happens during the burn-in, i.e. that, for any t, we are able to actually observe $\mathfrak{h}_t$.

In principle we admit the possibility of having different burn-in times for different units; it is furthermore assumed that the loss caused by choosing the durations $a_1, ..., a_n$ for the burn-in of the n units, when their life-lengths are $w_1, ..., w_n$ respectively, has the additive form

$$\tilde{l}(\mathbf{a}, \mathbf{w}) = \sum_{i=1}^{n} l(a_i, w_i) \tag{5.49}$$

where $l(a_i, w_i)$ is the loss caused by a burn-in of duration a_i for a single unit, when its life-length is w_i.

5.4.1 The case of i.i.d. lifetimes

Let us start with the case when $T_1, ..., T_n$ are independent.

In such a case, it is easily understood that the optimal burn-in time for each unit is a quantity which is independent of the burn-in times and of the behavior of the other units.

We then see that the burn-in problem for $T_1, ..., T_n$ reduces to a set of n independent burn-in problems for n single units.

In other words, we must minimize, with respect to $\mathbf{a} \in \mathbb{R}^n_+$, the expected value

$$\mathbb{E}\left[\tilde{l}(\mathbf{a}, \mathbf{T})\right] = \mathbb{E}\left[\sum_{i=1}^n l(a_i, T_i).\right]$$

Let us denote by a_i^* the optimal burn-in time for the unit i ($i = 1, ..., n$).

When $T_1, ..., T_n$ are independent and identically distributed, symmetry arguments show that it must be

$$a_1^* = a_2^* = ... a_n^* = a^*, \tag{5.50}$$

for some suitable value a^*. The value a^* then minimizes $\mathbb{E}\left[l(a; T_i).\right]$ and actually can be found by minimizing the expression in (5.7).

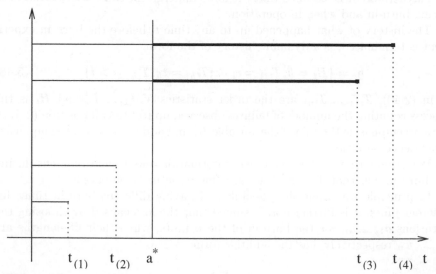

Figure 11. $n = 4, t_{(3)} - a^*, t_{(4)} - a^*$ are the operative lives of units surviving the deterministic burn-in time a^*.

Sometimes (as for instance in Definition 5.55 below) one may need to emphasize the dependence of a^* on the lifetime's survival function $\overline{F}$. In such cases, we shall use the symbol $a^*\left(\overline{F}\right)$.

Remark 5.47. We stress that, when $T_1, ..., T_n$ are i.i.d., the quantity a^*, the same optimal duration of burn-in for all the units, is a deterministic (nonnegative) quantity: it neither depends on the number of failures that can be observed at early times, nor on the values of the failure times that will be possibly observed; in other words, a^* is to be fixed at time 0. It is also remarkable that a^* does not depend on n, the total number of units to submit to burn-in.

In view of what was said before, we continue this subsection by concentrating attention on the burn-in problem for a single unit, the survival function of its lifetime T being denoted by $\overline{F}$.

Furthermore, we assume in particular that the loss function is as in (5.5); we recall that, in such a case, the quantity to be minimized is

$$\mathbb{E}[l\,(a,T)] = c + (C - c)\overline{F}(a) - (C + K)\overline{F}(a + \tau). \qquad (5.51)$$

Letting

$$\gamma = \frac{C - c}{C + K}, \qquad (5.52)$$

we can obtain:

$$a^*\left(\overline{F}\right) = 0 \;\; \text{if} \;\; \frac{f(a + \tau)}{f(a)} \geq \gamma, \forall a \geq 0$$

$$a^*\left(\overline{F}\right) = +\infty \;\; \text{if} \;\; \frac{f(a + \tau)}{f(a)} \leq \gamma, \forall a \geq 0$$

$$0 < a^*\left(\overline{F}\right) < \infty, \text{ if } \frac{f(\tau)}{f(0)} \leq \gamma \text{ and } \frac{f(a + \tau)}{f(a)} \geq \gamma \text{ for some } a > 0.$$

By differentiating (5.51) with respect to a, we can obtain more specifically

Proposition 5.48. *Under the condition that a value $\widehat{a} > 0$ exists such that*

$$\frac{f(a + \tau)}{f(a)} \leq \gamma \text{ for } a \leq \widehat{a}, \; \frac{f(a + \tau)}{f(a)} \geq \gamma \text{ for } a \geq \widehat{a} \qquad (5.53)$$

it is

$$a^*\left(\overline{F}\right) = \widehat{a}.$$

Note that the condition (5.53) certainly holds when, for the given value of τ and for some $\bar{a} > 0$, it is

$$\frac{f(\tau)}{f(0)} \leq \gamma, \quad \frac{f(\bar{a} + \tau)}{f(\bar{a})} > \gamma$$

and

$$\frac{f(a + \tau)}{f(a)} \text{ increasing.} \tag{5.54}$$

Remark 5.49. Let us consider the problem of optimally choosing the residual duration of burn-in for the residual lifetime $T - t$ given the survival data $T > t$. The corresponding density is then

$$f_{T-t}(x|T_1 > t) = \frac{f(t + x)}{\overline{F}(t)}$$

Assuming the condition (5.53) for $f(\cdot)$ implies that a similar condition also holds for $f_{T-t}(\cdot|T > t)$, by replacing $\hat{a}$ with $\hat{a} - t$. Then the optimal duration for the surviving unit is $\hat{a} - t$. This means that (if the unit does not fail before) we have to continue the burn-in up to reaching the age $\hat{a}$.

Remark 5.50. Obviously the condition (5.54) certainly holds for all $\tau \geq 0$ if $f(\cdot)$ is log-convex.

Example 5.51. Consider the case when T is conditionally exponential given a random variable Θ with density π_0:

$$f(t) = \int_0^\infty \theta \exp\{-\theta t\} \pi_0(\theta)\, d\theta. \tag{5.55}$$

$f(t)$ is then log-convex. If

$$\frac{f(\tau)}{f(0)} = \frac{\int_0^\infty \theta \exp\{-\theta \tau\} \pi_0(\theta)\, d\theta}{\mathbb{E}(\Theta)} \leq \gamma,$$

the optimal burn-in time is $a_{\pi_0}^*$ such that

$$\frac{f(a_{\pi_0}^* + \tau)}{f(a_{\pi_0}^*)} = \frac{\int_0^\infty \theta \exp\{-\theta(\tau + a_\pi^*)\} \pi_0(\theta)\, d\theta}{\int_0^\infty \theta \exp\{-\theta a_\pi^*\} \pi_0(\theta)\, d\theta} =$$

$$\mathbb{E}\left[\exp\{-\tau \Theta\}|T_1 = a_{\pi_0}^*\right] = \gamma. \tag{5.56}$$

In the following result, we consider two different densities f and g for the lifetime T, with corresponding survival functions $\overline{F}$ and $\overline{G}$, respectively. We assume that both f and g satisfy the condition (5.53). We want to compare the corresponding optimal burn-in times $a^*(\overline{F})$ and $a^*(\overline{G})$.

Proposition 5.52. *If* $f \leq_{lr} g$, *then* $a^* \left(\overline{F} \right) \geq a^* \left(\overline{G} \right)$.

Proof. By definition $f \leq_{lr} g$ implies

$$\frac{f(a + \tau)}{f(a)} \leq \frac{g(a + \tau)}{g(a)}.$$

Then if $\widehat{a}_f$ and $\widehat{a}_g$ are such that

$$\frac{f(a + \tau)}{f(a)} \leq \gamma \text{ for } a \leq \widehat{a}_f, \frac{f(a + \tau)}{f(a)} \geq \gamma \text{ for } a \geq \widehat{a}_f$$

and

$$\frac{g(a + \tau)}{g(a)} \leq \gamma \text{ for } a \leq \widehat{a}_g, \frac{g(a + \tau)}{g(a)} \geq \gamma \text{ for } a \geq \widehat{a}_g,$$

we can conclude that it must be $\widehat{a}_f \geq \widehat{a}_g$.   □

5.4.2   Dependence   and   optimal   adaptive   burn-in procedures

In this subsection and in the next one we present an introductory discussion and an informative description of the theme of optimal burn-in in the case of dependence among units' lifetimes, without developing formal arguments.

Let us then look at the case when $T_1, ..., T_n$ are dependent: we drop the assumption of stochastic independence but, in order to maintain the symmetry property of the problem, we keep the condition of exchangeability among the $T_i's$.

Due to the condition of dependence, a continuous process of learning takes place during the burn-in experiment; the consequence of this is twofold:

i) At any instant t (during the burn-in procedure) we should take into account the information, regarding residual lifetimes of the units, collected up to t, in order to decide about the residual duration of the procedure itself, for the surviving units. This means that only "adaptive" procedures should be considered.

ii) When planning, at time 0, the burn-in procedure, we should optimize also taking into account the expected amount of information that we could collect during the experiment.

From a technical point of view, choosing an adaptive burn-in procedure means to choose a stopping time $\mathcal{T}$, adapted to the flow of information described by $\{\mathfrak{h}_t\}_{t \geq 0}$ as in (5.48) i.e. a non-negative random variable $\mathcal{T}$ such that, for any $t \geq 0$, we can establish with certainty if the event $\{\mathcal{T} \leq t\}$ or $\{\mathcal{T} > t\}$ is true only based on the information contained in $\mathfrak{h}_t$.

In practice, in order to construct an adaptive procedure we have to reconsider the residual duration of the procedure only at the instants when possible failures arrive and to take into account the information, collected up to those instants (we have then a "multistage" decision problem).

This is an intuitive fact, that could, nevertheless, also be proved in rigorous terms by using Theorem T 33 in Bremaud (1981, p. 308).

For this reason we consider sequential burn-in procedures of the following type:

- At the instant 0, when all the n units are working and are new, we start the burn-in procedure, planning to stop it at a deterministic time $a^{(n)}$, if no failure will be observed in between.

- If $T_{(1)} = t_1 < a^{(n)}$, at t_1 we reconsider the residual duration of burn-in for the remaining $(n-1)$ units. This will be a non-negative quantity which takes into account the past information (i.e. $\{T_{(1)} = t_1\}$) and that will be denoted by $a^{(n-1)}(t_1)$; generally , $a^{(n-1)}(t_1)$ will be different from $\left(a^{(n)} - t_1\right)$, as initially planned; notice that, when determining $a^{(n-1)}(t_1)$, the $(n-1)$ surviving units will have an age t_1.

- If one of the units fails before $t_1 + a^{(n-1)}(t_1)$ (i.e. if $T_{(2)} = t_2$ with $t_2 < t_1 + a^{(n-1)}(t_1)$), we fix the residual duration of burn-in equal to a non-negative quantity, denoted by $a^{(n-2)}(t_1, t_2)$ and so on.

A generic adaptive procedure of this kind will be denoted by the symbol $\mathbb{S}$.

When we need to avoid ambiguities and to emphasize that we are dealing with a specific procedure $\mathbb{S}$ we shall write $a_{\mathbb{S}}^{(n)}, a_{\mathbb{S}}^{(n-1)}(t_1), ..., a_{\mathbb{S}}^{(1)}(t_1, .., t_{n-1})$.

Then $\mathbb{S}$ is determined by the sequence

$$a_{\mathbb{S}}^{(n)}, a_{\mathbb{S}}^{(n-1)}(t_1), a_{\mathbb{S}}^{(n-2)}(t_1, t_2), ..., a_{\mathbb{S}}^{(1)}(t_1, .., t_{n-1}). \tag{5.57}$$

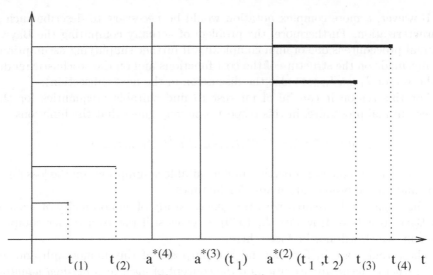

Figure 12. $n = 4, t_3 - a^{*(2)}(t_1, t_2), t_4 - a^{*(2)}(t_1, t_2)$ are the operative lives of units surviving the burn-in time $a^{*(2)}(t_1, t_2)$.

Not any adaptive procedure $\mathbb{S}$ will be Bayes optimal with respect to the preassigned loss function $\tilde{l}$.

A Bayes optimal procedure $\mathbb{S}$ is one that minimizes the overall expected cost over all possible adaptive procedures of the form in (5.57). Such a procedure can be indicated by

$$\mathbb{S}^* \equiv [a^{(n)*}, a^{(n-1)*}(t_1), a^{(n-2)*}(t_1, t_2), ..., a^{(1)*}(t_1, .., t_{n-1})].$$

Remark 5.53. Consider the case when $T_1, ..., T_n$ are i.i.d.:

$$\overline{F}^{(n)}(t_1, ..., t_n) = \overline{F}^{(1)}(t_1) \cdot ... \cdot \overline{F}^{(1)}(t_n).$$

Due to identical distribution the residual burn-in time must be the same for all the units surviving at an instant $s > 0$. Due to independence such a residual burn-in time must be independent of the failure times observed past to s and coincides with the optimal solution of the burn-in problem for a single unit of age s.

We can expect that it is (see Remark 5.49 for special cases)

$$a^{(n)*} = a^*(\overline{F}^{(1)}), a^{(n-1)*}(t_1) = a^*(\overline{F}^{(1)}) - t_1, ..., a^{*(1)}(t_1, .., t_{n-1}) = a^*(\overline{F}^{(1)}) - t_{n-1}.$$

See Herberts and Jensen (1999) for a formal treatment.

Some discussions about the characterization of $\mathbb{S}^*$ by means of dynamic programming and about other aspects of the Bayes-optimal burn-in procedure, have been presented by Spizzichino, 1991 and Runggaldier, 1993.

However, a more complex notation would be necessary to describe such a characterization. Furthermore, the problem of actually computing the Bayes-optimal procedure is one of high complexity, if further simplifying assumptions are not made on the structure of the cost functions and on the stochastic model of the vector $T_1, ..., T_n$ (see also the discussion in the next subsection).

For this reason it can be of interest to find suitable inequalities for the Bayes-optimal procedure; in this respect, one may guess that the functions

$$a^{(n-1)*}(t_1), a^{(n-2)*}(t_1, t_2), ..., a^{(1)*}(t_1, .., t_{n-1})$$

have some monotonicity properties, under suitable assumptions on the loss functions and on the probability model for lifetimes.

This topic can be related with the general study of monotonicity of sequential Bayes test (see Brown et al., 1979); it gives still rise to a rather complex problem, which will not be faced, here.

These topics are in fact beyond the purposes of this monograph and we shall rather concentrate attention on the concept of *open-loop optimal adaptive procedure*; this will be treated in Subsection 5.4.4.

Denoting by $\widehat{\mathbb{S}}^*$ the open-loop adaptive optimal burn-in procedure, which will be defined therein, we shall see, instead, that some inequalities for the functions

$$a_{\widehat{\mathbb{S}}^*}^{(n-1)}(t_1), a_{\widehat{\mathbb{S}}^*}^{(n-2)}(t_1, t_2), ..., a_{\widehat{\mathbb{S}}^*}^{(1)}(t_1, .., t_{n-1})$$

may be obtained as an application of the previous Proposition 5.52.

5.4.3   Burn-in, optimal stopping, monotonicity, and Markovianity

Different models and formulations exist of burn-in problems and of their optimal solutions. It is a general fact however that burn-in problems can be seen as optimal stopping problems for suitable stochastic processes (see Aven and Jensen, 1999 and references cited therein).

High complexity of optimal stopping problems compels to the search for special structures which can be exploited to reduce complexity or to *a priori* prove qualitative properties that the optimal solutions should have. Special structures often involve monotonicity properties of the cost functions and/or of the involved stochastic processes.

In Aven and Jensen (1999), an approach is illustrated, based on the notion of *semimartingale* and on a specific notion of monotonicity. The approach there permits one to deal with quite general models, where, in particular, the processes to be stopped are not necessarily Markov.

Indeed there are problems of interest where Markovianity is not a realistic assumption. When, on the other hand, the process to be stopped actually has such a property, the theory of optimal stopping of Markov processes is available (see Shiryaev, 1978).

Being interested in the study of models admitting dependence, we fixed attention on the condition of exchangeability. For exchangeable lifetimes $T_1, ..., T_n$ a useful connection with the property of Markovianity is sketched in what follows.

For any $t > 0$, consider the $(n + 1)$-dimensional random vector defined by

$$\mathcal{H}_t \equiv (H_t, T_{(1)} \wedge t, ..., T_{(n)} \wedge t);$$

The first coordinate H_t counts the number of failures observed up to t, then it is

$$T_{(1)} = T_{(1)} \wedge t, ..., T_{(H_t)} = T_{(H_t)} \wedge t; \; T_{(H_t+1)} > t.$$

When $T_{(H_t)} < t$ (which in regular cases happens with probability one) $\mathcal{H}_t$ contains then a redundant information; however it can be convenient for different reasons to consider the redundant description $\mathcal{H}_t$ in place of $(T_{(1)} \wedge t, ..., T_{(n)} \wedge t)$.

The process $\{\mathcal{H}_t\}_{t \geq 0}$ has the following important properties:

i) The knowledge of the value of $\mathcal{H}_t$ is equivalent to the observation of the event $\mathfrak{h}_t$ in (5.48).

ii) $\{\mathcal{H}_t\}_{t \geq 0}$ is a Markov process.

iii) Any adaptive burn-in procedure for the lifetimes $T_1, ..., T_n$ can be seen as a stopping time for $\{\mathcal{H}_t\}_{t \geq 0}$ (for the definition of stopping time, see e.g. Bremaud, 1981, or Shiryaev, 1978).

iv) We can assume that the cost function in a burn-in problem is such that the cost of stopping burn-in at a generic instant t is only a function of $\mathcal{H}_t$.

Finding a Bayes optimal adaptive burn-in procedure then becomes an optimal stopping problem for the process $\{\mathcal{H}_t\}_{t \geq 0}$, i.e. for a $(n + 1)$ dimensional Markov process. The theory of optimal stopping of Markov processes can be applied to characterize the optimal Bayes burn-in procedures and to study their properties.

However, even for Markov processes, analytic procedures to find the solution of optimal stopping problems are seldom available; also, numerical procedures can be difficult due to great complexity involved in the computations.

Since the computational complexity increases with the dimension of the process which is to be stopped, a possibly useful idea is the following: try to transform an optimal stopping problem of $\{\mathcal{H}_t\}_{t \geq 0}$ into one for another process of reduced dimension.

The notion of dynamic sufficiency (Definition 2.60) can be of help in this respect.

First note that the set $\widehat{\mathcal{G}}$ defined in (2.75) coincides with the state-space of the process $\{\mathcal{H}_t\}_{t \geq 0}$. Then a function q defined on $\widehat{\mathcal{G}}$ gives rise to a new stochastic process $\{Q_t\}_{t \geq 0}$, where

$$Q_t \equiv q(\mathcal{H}_t)$$

It can be seen that, under the assumption that q is a dynamic sufficient statistic and under some additional technical condition, Q_t is also a Markov process.

Let us then think of models for $(T_1, ..., T_n)$ for which a low-dimension dynamic sufficient statistic q exists so that $Q_t = q(\mathcal{H}_t)$ is a Markov process, and the dimension of its state-space is smaller than n.

Sometimes, in a burn-in problem, the loss function is such that the cost of stopping the burn-in at t is only a function of Q_t.

In such cases the burn-in problem can become an optimal stopping problem of the process Q_t.

Example 5.54. As we saw in Subsection 2.3.4, a relevant example of dynamic sufficiency is the following: when $T_1, ..., T_n$ admit a Schur-constant joint density, a dynamic sufficient statistic can be found in the pair

$$Z_t \equiv (H_t, Y_t) \tag{5.58}$$

where Y_t is the TTT process defined in Section 2.4:

$$Y_t = \sum_{i=1}^{n} \left(T_{(i)} \wedge t \right).$$

It can easily be seen that, when $f^{(n)}$ is Schur-constant, $\{Z_t\}_{t \geq 0}$ in (5.58) is a (two-dimensional) Markov process, which moreover turns out to have the property of *stochastic monotonicity* when $T_1, ..., T_n$ are conditionally independent, exponentially distributed (see Caramellino and Spizzichino, 1996).

When the cost for conducting a burn-in up to time t depends only on the value of Z_t, one can actually reduce the (Bayes) optimal burn-in problem to an optimal stopping problem for the process $\{Z_t\}_{t \geq 0}$ (Costantini and Spizzichino, 1997).

Under the condition that $T_1, ..., T_n$ are conditionally independent and exponentially distributed, such an optimal stopping problem has furthermore a special structure, based on some stochastic comparisons in the lr sense. Such a structure derives from the mentioned stochastic monotonicity of $\{Z_t\}_{t \geq 0}$ and from a suitable monotonicity property of the considered cost function; it can be exploited to find the solution in a rather explicit form. This is then a case where the Bayes optimal burn-in procedure can actually be computed.

5.4.4 Stochastic orderings and open-loop optimal adaptive burn-in procedures

In this subsection we give the definition of open-loop feedback optimal burn-in, then we show some related role of the arguments discussed in the previous sections.

The general concept of open-loop feedback optimality (OLFO) has been introduced within the theory of optimal control, where OLFO solutions of control problems are compared with *closed-loop feedback optimal solutions* (the reader can find in the paper by Runggaldier (1993a) a review and an essential list of references on optimal control theory).

Closed-loop feedback optimal solutions are those corresponding to Bayes optimality and, as mentioned, the problem of their computation is often a very complex one.

OLFO solutions are more easily computable. Actually they are suboptimal according to the Bayesian paradigm of minimizing the expected cost; however they are generally reasonably good and not as complex to be computed. In any case they are based on a still Bayesian (even if suboptimal) logic, as will be seen later for our specific case of interest.

It is to be noted that the problem of finding an optimal burn-in procedure, substantially being a special problem of optimal stopping, can be seen as a very particular problem in the theory of optimal control.

Then the concept of open-loop feedback optimality can be applied to the burn-in problem (see Runggaldier, 1993b).

We are going to define what is meant by the term *OLFO* burn-in procedure, next. The term *adaptive*, which is more familiar to statisticians, will be sometimes used in the following in place of *feedback*.

Let us, as usual, denote by $\overline{F}^{(1)}$ the one-dimensional marginal distribution of $T_1, ..., T_n$.

For a dynamic history $\mathfrak{h}_t$ as in (5.48), let moreover $\overline{F}_t^{(1)} (\cdot|h; t_1, ..., t_h)$ denote the (one-dimensional) conditional distribution, given $\mathfrak{h}_t$, of the residual lifetimes of the $(n - h)$ units (of age t) which are still surviving at t.

Definition 5.55. The open-loop optimal adaptive burn-in procedure is the adaptive burn-in procedure $\widehat{\mathbb{S}}^*$ defined by the positions

$$a_{\widehat{\mathbb{S}}^*}^{(n)} = a^*(\overline{F}^{(1)}), a_{\widehat{\mathbb{S}}^*}^{(n-1)}(t_1) = a^* \left(\overline{F}_{t_1}^{(1)} (\cdot|1; t_1) \right),$$

$$a_{\widehat{\mathbb{S}}^*}^{(n-2)}(t_1, t_2) = a^* \left(\overline{F}_{t_2}^{(1)} (\cdot|2; t_1, t_2) \right), ...$$

$$..., a_{\widehat{\mathbb{S}}^*}^{(1)}(t_1, .., t_{n-1}) = a^* \left(\overline{F}_{t_{n-1}}^{(1)} (\cdot|n - 1; t_1, ..., t_{n-1}) \right). \tag{5.59}$$

Remark 5.56. In words, the essence of an OLFO burn-in procedure $\widehat{\mathbb{S}}^*$ can be explained as follows: at time 0, we fix the burn-in duration equal to

$$a_{\widehat{\mathbb{S}}^*}^{(n)} = a^*(\overline{F}^{(1)}).$$

This means the following: initially we take into account that the common marginal survival function of $T_1, ..., T_n$ is $\overline{F}^{(1)}$, but we ignore the structure of dependence among those lifetimes, i.e. we behave as if $T_1, ..., T_n$ were independent.

If no failure is observed before the time $a^*(\overline{F}^{(1)})$, then, at that time, we deliver all the units to operation.

If, on the contrary, $T_{(1)} < a^*(\overline{F}^{(1)})$, we reconsider, at $T_{(1)}$, the duration of the residual burn-in for surviving units. We still go on as if the residual lifetimes were independent; however we take into account that the updated (one-dimensional) survival function for them is now $\overline{F}_{t_1}^{(1)}(\cdot|1; t_1)$ and then we set

$$a_{\widehat{\mathbb{S}}*}^{(n-1)}(t_1) = a^* \left(\overline{F}_{t_1}^{(1)}(\cdot|1; t_1) \right)$$

and so on.

Remark 5.57. When $T_1, ..., T_n$ are i.i.d., with a marginal survival function $\overline{F}^{(1)}$, the OLFO burn-in procedure does obviously coincide with the closed-loop feedback optimal procedure and it is

$$a_{\widehat{\mathbb{S}}*}^{(n)} = a^* \left(\overline{F}^{(1)} \right), \quad a_{\widehat{\mathbb{S}}*}^{(n-1)}(t_1) = a^* \left(\overline{F}^{(1)} \right) - t_1, ...$$

$$..., a_{\widehat{\mathbb{S}}*}^{(1)}(t_1, .., t_{n-1}) = a^* \left(\overline{F}^{(1)} \right) - t_{n-1}$$

(compare also with Remark 5.53).

Example 5.58. (OLFO burn-in of conditionally exponential units). Let $T_1, ..., T_n$ be conditionally independent given $\{\Theta = \theta\}$, with density $f(t|\theta) = \theta \exp\{-\theta t\}$. Denoting by $\pi_0(\cdot)$ the prior density of Θ, we then have

$$f^{(n)}(t_1, ..., t_n) = \int_0^\infty \theta^n \exp\{-\theta \sum_{i+1}^n t_i\} \pi_0(\theta) \, d\theta$$

We assume the loss function as in (5.5).

Since the one-dimensional marginal density is as in (5.55), we have that

$$a_{\widehat{\mathbb{S}}*}^{(n)} = a_\pi^*$$

where a_π^* is given as in (5.56).

If $T_{(1)} = t_1 < a_{\widehat{\mathbb{S}}*}^{(n)}$, we must continue, after t_1 the burn-in procedure for an extra time $a_{\widehat{\mathbb{S}}*}^{(n-1)}(t_1)$. Now note that, conditionally on the observation

$$\mathfrak{h}_{t_1} = \{T_{(1)} = t_1, T_{(2)} > t_1\},$$

the conditional density of Θ becomes

$$\pi\left(\theta|\mathfrak{h}_{t_1}\right) = \frac{\theta\exp\{-\theta n t_1\}\pi_0\left(\theta\right)}{\int_0^\infty \theta\exp\{-\theta n t_1\}\pi_0\left(\theta\right)d\theta}.$$

whence the conditional density of the residual lifetimes is

$$f\left(t|1;t_1\right) = \frac{\int_0^\infty \theta^2 \exp\{-\theta(t+n t_1)\}\pi_0\left(\theta\right)d\theta}{\int_0^\infty \theta\exp\{-\theta n t_1\}\pi_0\left(\theta\right)d\theta};$$

furthermore

$$a_{\widehat{\mathbb{S}}_*}^{(n-1)}\left(t_1\right) = a^*_{\pi\left(\theta|\mathfrak{h}_{t_1}\right)}.$$

In a similar way we can obtain

$$a_{\widehat{\mathbb{S}}_*}^{(n-2)}\left(t_1,t_2\right), ..., a_{\widehat{\mathbb{S}}_*}^{(1)}\left(t_1,..,t_{n-1}\right).$$

Notice that $a_{\widehat{\mathbb{S}}_*}^{(n-h)}\left(t_1,..,t_h\right)$ depends only on h and on the observed total time on test statistic $\sum_{i=1}^h t_i + (n-h)t_h$.

Now we proceed to briefly discuss how the Proposition 5.52 can be applied to obtain useful properties of an OLFO burn-in procedure, when the loss function is as in (5.5).

From the very definition, we see that an OLFO burn-in procedure depends on $\overline{F}^{(1)}$ and on the (one-dimensional) conditional distributions

$$\overline{F}_{t_1}^{(1)}\left(\cdot|1;t_1\right), \overline{F}_{t_2}^{(1)}\left(\cdot|2;t_1,t_2\right), ..., \overline{F}_{t_{n-1}}^{(1)}\left(\cdot|n-1;t_1,t_2,...,t_{n-1}\right)$$

for the residual lifetimes of the units, respectively surviving after the progressive failure times $T_{(1)},...,T_{(n-1)}$.

On the other hand, in Sections 5.2 and 5.3, we obtained a sample of results leading to establish stochastic comparisons, in the lr sense, between two (one-dimensional) conditional distributions for residual lifetimes given two different dynamic histories.

By Proposition 5.52 then, such stochastic comparisons can be used to possibly obtain appropriate monotonicity properties of an OLFO procedure of burn-in.

In particular, as a consequence of (5.23) and Proposition 5.52, we have

Lemma 5.59. *Let the joint density $f^{(n)}$ of $T_1,...,T_n$ be MTP_2 and let the conditional survival functions $\overline{F}_t^{(1)}\left(\cdot|h;t_1,...,t_h\right)$ admit densities satisfying condition (5.53),$\forall h = 1,2,...,n-1, 0 \leq t_1 \leq ... \leq t_h$. Then, for $0 \leq t_1 \leq t_2 \leq ... \leq t_h < t_{h+1}$ such that*

$$t_{h+1} < a_{\widehat{\mathbb{S}}_*}^{(n-h)}\left(t_1,..,t_h\right),$$

we have that

$$a_{\widehat{\mathbb{S}}*}^{(n-h)}(t_1,..,t_h) \text{ is a decreasing function of } t_1,...,t_h.$$

Consider now specifically the case of conditional independence, given a scalar parameter Θ, where the joint density $f^{(n)}$ is of the form (5.40), with $g(t|\theta) > 0, \forall t > 0$; we see that, under suitable additional conditions on $g(t|\theta)$, we have a result stronger than Lemma 5.59.

Proposition 5.60. *Let* $\log g(t|\theta)$ *be a convex function of* t *and* $\frac{\partial}{\partial\theta} \log g(t|\theta)$ *be a convex, decreasing function of* $t, \forall\theta$. *Then*

$$a_{\widehat{\mathbb{S}}*}^{(n-h)}(t'_1,..,t'_{h-1},t_h) \geq a_{\widehat{\mathbb{S}}*}^{(n-h)}(t_1,..,t_{h-1},t_h)$$

whenever $\mathbf{t}' \preceq \mathbf{t}$.

Proof. Let π be any initial density of Θ and consider the one-dimensional predictive density defined by

$$f(t) = \int_L g(t|\theta)\pi(\theta)\,d\theta. \tag{5.60}$$

We can write, for $x > 0$,

$$\frac{f(t+x)}{f(t)} = \frac{\int_L g(t+x|\theta)\pi(\theta)\,d\theta}{\int_L g(t|\theta)\pi(\theta)\,d\theta} =$$

$$\frac{\int_L \frac{g(t+x|\theta)}{g(t|\theta)}g(t|\theta)\pi(\theta)\,d\theta}{\int_L g(t|\theta)\pi(\theta)\,d\theta} = \int_L \frac{g(t+x|\theta)}{g(t|\theta)}\pi(\theta|T=t)\,d\theta,$$

by setting

$$\pi(\theta|T=t) = \frac{g(t|\theta)\pi(\theta)}{\int_L g(t|\theta)\pi(\theta)\,d\theta}.$$

$\frac{\partial}{\partial\theta} \log g(t|\theta)$ being a decreasing function of $t, \forall\theta$, it follows that $\frac{g(t+x|\theta)}{g(t|\theta)}$ is a decreasing function of θ and then for $t' > t$,

$$\pi(\theta|T=t') \leq_{lr} \pi(\theta|T=t)$$

$$\int_L \frac{g(t'+x|\theta)}{g(t'|\theta)}\pi(\theta|T=t')\,d\theta \geq \int_L \frac{g(t'+x|\theta)}{g(t'|\theta)}\pi(\theta|T=t)\,d\theta.$$

On the other hand, $\log g(t|\theta)$ being a convex function of t means that

$$\frac{g(t'+x|\theta)}{g(t'|\theta)} \geq \frac{g(t+x|\theta)}{g(t|\theta)}, \forall \theta,$$

whence

$$\int_L \frac{g(t'+x|\theta)}{g(t'|\theta)} \pi\,(\theta|T=t)\,d\theta \geq \int_L \frac{g(t+x|\theta)}{g(t|\theta)} \pi\,(\theta|T=t)\,d\theta.$$

Then

$$\frac{f\,(t'+x)}{f\,(t')} = \int_L \frac{g(t'+x|\theta)}{g(t'|\theta)} \pi\,(\theta|T=t')\,d\theta \geq$$

$$\int_L \frac{g(t+x|\theta)}{g(t|\theta)} \pi\,(\theta|T=t)\,d\theta = \frac{f\,(t+x)}{f\,(t)}$$

and we can then claim that a density $f\,(t)$ of the form (5.60) is log-convex, for any initial density π.

Now a conditional survival function of type $\overline{F}_t^{(1)}\,(\cdot|h;t_1,...,t_h)$ admits a density given by

$$f_t^{(1)}\,(u|h;t_1,...,t_h) = \frac{f^{(1)}\,(t+u|h;t_1,...,t_h)}{\overline{F}^{(1)}\,(t|h;t_1,...,t_h)}$$

$$= \frac{\int_L g(t+u|\theta)\pi\,(\theta|h;t_1,...,t_h)\,d\theta}{\int_L \overline{G}(t|\theta)\pi\,(\theta|h;t_1,...,t_h)\,d\theta}, \qquad (5.61)$$

whence

$$\frac{f_t^{(1)}\,(u+x|h;t_1,...,t_h)}{f_t^{(1)}\,(u|h;t_1,...,t_h)} = \frac{\int_L g(t+u+x|\theta)\pi\,(\theta|h;t_1,...,t_h)\,d\theta}{\int_L g(t+u|\theta)\pi\,(\theta|h;t_1,...,t_h)\,d\theta}$$

By the arguments above we see then that $f_{t_h}^{(1)}\,(\cdot|h;t_1,...,t_{h-1},t_h)$ is a log-convex density, for any dynamic history $\mathfrak{h}_t$ as in (5.48).

By using Proposition 5.52, we can now conclude the proof by showing that

$$f_{t_h}^{(1)}\,(\cdot|h;t_1,...,t_{h-1},t_h) \geq_{lr} f_{t_h}^{(1)}\,(\cdot|h;t_1',...,t_{h-1}',t_h)\,.$$

In view of Equation (5.61), and since an analogous expression holds for

$$f_{t_h}^{(1)}\,(\cdot|h;t_1',...,t_{h-1}',t_h)\,,$$

the above comparison is achieved by showing that

$$\pi(\theta|T_1 = t_1, ..., T_{h-1} = t'_{h-1}, T_h = t_h) \leq_{lr}$$

$$\pi(\theta|T_1 = t'_1, ..., T_{h-1} = t'_{h-1}, T_h = t_h).$$

Now

$$\frac{\pi(\theta|T_1 = t_1, ..., T_{h-1} = t_{h-1}, T_h = t_h)}{\pi(\theta|T_1 = t'_1, ..., T_{h-1} = t'_{h-1}, T_h = t_h)} = \frac{\prod_{j=1}^{h-1} g(t_j|\theta)}{\prod_{j=1}^{h-1} g(t'_j|\theta)};$$

$\frac{\partial}{\partial\theta} \log g(t|\theta)$ being convex, it follows

$$\frac{\partial}{\partial\theta} \frac{\pi(\theta|T_1 = t_1, ..., T_h = t_h)}{\pi(\theta|T_1 = t'_1, ..., T_h = t'_h)} \leq 0.$$

$\square$

Proposition 5.61. *Let the conditions of Proposition 5.60 hold and let* $t_1, ..., t_{h-1}$, $\tilde{t}_1, ..., \tilde{t}_{h-1}$ *be such that*

$$t_1 \leq ... \leq t_{h-1}, \tilde{t}_1 \leq ... \leq \tilde{t}_{h-1}$$

$$t_1 \geq \tilde{t}_1, t_1 + t_2 \geq \tilde{t}_1 + \tilde{t}_2, ..., \sum_{i=1}^{h-1} t_i \geq \sum_{i=1}^{h-1} \tilde{t}_i. \tag{5.62}$$

Then

$$a_{\widehat{\mathbb{S}}*}^{(n-h)}(\tilde{t}_1, .., \tilde{t}_{h-1}, t_h) \geq a_{\widehat{\mathbb{S}}*}^{(n-h)}(t_1, .., t_{h-1}, t_h). \tag{5.63}$$

Proof. By the conditions (5.62), we can find a vector $\mathbf{t}' \equiv \left(t'_1, ..., t'_{h-1}\right)$ such that

$$\mathbf{t}' \leq \mathbf{t}, \mathbf{t}' \prec \tilde{\mathbf{t}}.$$

Then, by using Lemma 5.59 and Proposition 5.60, we have

$$a_{\widehat{\mathbb{S}}*}^{(n-h)}(\tilde{t}_1, ..., \tilde{t}_h) \geq a_{\widehat{\mathbb{S}}*}^{(n-h)}(t'_1, .., t'_h) \geq a_{\widehat{\mathbb{S}}*}^{(n-h)}(t_1, .., t_h).$$

$\square$

Notice that the inequality (5.62) obviously is a weaker condition than $\tilde{\mathbf{t}} \leq \mathbf{t}$.

Remark 5.62. Suppose that we collected two different sets of failure data $(t_1, .., t_{h-1})$ and $(t'_1, .., t'_{h-1})$, until time s, for two different sets of units $U_1, .., U_n$ and $U'_1, .., U'_n$, respectively.

The condition $\mathbf{t}' \preceq \mathbf{t}$ can be seen as an indication that the units $U'_1, .., U'_n$ still are (more than it can happen for $U_1, .., U_n$) in the period of infantile mortality, at time s.

This argument supplies a heuristic interpretation of the result in Proposition 5.61.

5.5   Exercises

Exercise 5.63. Let Q, c and $K(t)$, in Example 5.5, be given. Apply the formula (5.8) to find the optimal time of preventive replacement a^* for a unit whose lifetime has density of the form

$$f^{(1)}(t) = \int_0^\infty \theta \rho(t) \exp\{-\theta R(t)\} \pi_0(\theta)\, d\theta, \quad \rho(t) = \frac{d}{dt} R(t)$$

Deduce that $\rho(t)$ decreasing implies $a^* = 0$ or $a^* = \infty$.

Exercise 5.64. In Example 5.5, take now $K(t) = q \cdot t$, for some $q > 0$ and consider the optimal times of replacement a_1^*, a_2^* for two units U_1 and U_2, whose lifetimes are IFR and have densities $f_1^{(1)}$ and $f_2^{(1)}$ respectively. Check that $f_1^{(1)} \leq_{lr} f_2^{(1)}$ implies $a_1^* \leq a_2^*$.

Exercise 5.65. In Example 5.8, consider the case of a proportional hazard model for $T_1, ..., T_n, T$, defined by a joint density of the form (5.42), with the special choice

$$\pi_0(\theta) = \frac{\beta^\alpha}{\Gamma(\alpha)} \theta^{\alpha-1} \exp\{-\beta\theta\},$$

for given $\alpha > 0, \beta > 0$. Write explicitly $a^*(t_1, ..., t_n)$ in terms of $\rho(\cdot)$. Check then directly that $a^*(t_1, ..., t_n)$ is Schur-concave or Schur-convex according to $\rho(\cdot)$ being decreasing or increasing.

Exercise 5.66. For the same model as in Exercise 5.65 find the solution of the predictive spare parts problem in Example 5.16.

Exercise 5.67. Check that the condition (5.33) holds for the joint survival function considered in Example 5.34.

Hint: $\overline{H}$ is decreasing and $\max_{1 \leq i \leq n-1} r_i$ is a Schur-convex function of $(r_1, ..., r_{n-1})$.

Exercise 5.68. Prove Proposition 5.40.

Hint: split the proof for the two cases when $W(t_1', ..., t_{n-1}', t) \geq 0$ and $W(t_1', ..., t_{n-1}', t) < 0$ and notice that, for $\mathbf{t}$ such that $\frac{\partial}{\partial t} f^{(n)}(t_1, ..., t_{n-1}, t) \geq 0$, Schur-convexity of $\frac{\partial}{\partial t} f^{(n)}(t_1, ..., t_{n-1}, t)$ and the condition $\mathbf{t}' \preceq \mathbf{t}$ imply that only two cases are possible:

$$a)\, 0 \leq \frac{\partial}{\partial t} f^{(n)}(t_1', ..., t_{n-1}', t) \leq \frac{\partial}{\partial t} f^{(n)}(t_1, ..., t_{n-1}, t)$$

$$b)\, \frac{\partial}{\partial t} f^{(n)}(t_1', ..., t_{n-1}', t) \leq 0 \leq \frac{\partial}{\partial t} f^{(n)}(t_1, ..., t_{n-1}, t).$$

Exercise 5.69. Consider the proportional hazard model for conditionally i.i.d. lifetimes $T_1, ..., T_n$ and find sufficient conditions on $\rho(\cdot)$ which imply the assumptions used decreasing, log-convex implies that the corresponding one-dimensional density $f^{(1)}$ is log-convex, for any choice of the prior density π_0.

For the case when π_0 is a gamma density (i.e. the same model as in Exercise 5.65) give an explicit expression to the ratio $\frac{f^{(1)}(t+\tau)}{f^{(1)}(t)}$.

By applying Proposition 5.52, find, for the case above, an expression for the corresponding optimal solution of the one-dimensional burn-in problem, with cost function of the form as in (5.5).

Consider now the OLFO procedure. Using the expression found above and replacing π_0 with the conditional density of Θ given an observation of the type $D = \{T_{(1)} = t_1, ..., T_{(h)} = t_h\}$, show directly, for the present special case, the validity of the inequality (5.62) for $(t_1, ..., t_{h-1})$ and $(\tilde{t}_1, ..., \tilde{t}_{h-1})$ as in Proposition 5.61.

Exercise 5.70. Consider the problem of optimal time replacement for units with exchangeable lifetimes. Similarly to the case of the burn-in problem, we can argue that, in the case of stochastic dependence, the optimal replacement policy should be adaptive, i.e. the decision as to replacing a unit of age r at a time t, should depend on the history of failures observed up to t. For the case of a loss function specified by the position $K(t) = q \cdot t$, formulate a definition of "open loop optimal" time replacement policy and explain the possible use of the result in Exercise 5.64.

5.6 Bibliography

Aven, T. and Jensen, U. (1999). *Stochastic models in reliability.* Applications of Mathematics, 41. Springer-Verlag, New York.

Barlow, R.E. (1998). *Engineering reliability.* Society for Industrial and Applied Mathematics (SIAM), Philadelphia.

Barlow, R. E. and Proschan, F. (1965). *Mathematical theory of reliability.* John Wiley & Sons, New York.

Barlow, R. E. and Proschan, F. (1975). *Statistical theory of reliability. Probability models.* Holt, Rinehart and Wiston, New York.

Barlow, R. E. and Proschan, F. (1988). Life Distribution Models and Incomplete Data. In *Handbook of Statistics,* Vol. 7, 225–250. North-Holland/Elsevier, Amsterdam, New York.

Barlow, R.E. and Zhang, X. (1987). Bayesian analysis of inspection sampling procedures discussed by Deming. *J. Stat. Plan. Inf.* 16, no. 3, 285–296.

Barlow, R. E. , Clarotti, C. A. and Spizzichino, F. (Eds) (1990). *Reliability and decision making.* Proceedings of the conference held at the University of Siena, Siena, October 15–26. Chapman & Hall, London, 1993.

Berg, M. (1997). Performance comparisons for maintained items. Stochastic models of reliability. *Math. Meth. Oper. Res.* 45, no. 3, 377–385.

Berger, J. (1985). *Statistical decision theory and Bayesian analysis.* Second edition. Springer-Verlag, New York-Berlin.

Bertsekas, D. (1976). *Dynamic programming and stochastic control.* Mathematics in Science and Engineering, 125. Academic Press, New York.

Billingsley, P. (1995). *Probability and measure.* Third edition. John Wiley & Sons, New York.

Block, H.W. and Savits, T. H.(1997). Burn-in. *Statistical Science.* 12, No. 1, 1-19.

Block, H.W. and Savits, T. H. (1994). Comparison of maintenance policies. In *Stochastic Orders and Their Applications* (Shaked and Shamthikumar, Eds). Academic Press, London.

Bremaud, P. (1981). *Point Processes and Queues. Martingale dynamics.* Springer Verlag, New York-Berlin.

Brown, L.D., Cohen, A. and Strawderman, W.E (1979). Monotonicity of Bayes Sequential Tests. *Ann. Math. Stat.*, 7, 1222-1230.

Caramellino, L. and Spizzichino, F. (1996). WBF property and stochastic monotonicity of the Markov Process associated to Schur-constant survival functions. *J. Multiv. Anal.*, 56, 153-163.

Clarotti, C.A. and Spizzichino, F. (1990). Bayes burn-in decision procedures. *Probab. Engrg. Inform. Sci.*, 4, 437-445.

Clarotti, C.A. and Spizzichino, F. (1996). Bayes predictive design of scram systems: the related mathematical and philosophical implications. *IEEE Trans. on Rel.*,45, 485-490.

Costantini, C. and Pasqualucci, D. (1998). Monotonicity of Bayes sequential tests for multidimensional and censored observations. *J. Statist. Plann. Inference,*75 , no. 1, 117–131.

Costantini, C. and Spizzichino, F. (1997). Explicit solution of an optimal stopping problem: the burn-in of conditionally exponential components. *J. Appl. Prob*, 34, 267-282.

Deming, W. E. (1982). *Quality, Productivity and Competitive Position* (M.I.T., Center for Advanced Engineering Study, Cambridge, Ma).

De Groot, M. H. (1970). *Optimal Statistical Decisions.* McGraw-Hill, New York.

Ferguson, T.S. (1967). *Mathematical statistics: A decision theoretic approach.* Academic Press, New York-London.

Gertsbakh, I. (2000). *Reliability theory with applications to preventive maintenance.* Springer-Verlag, New York.

Herberts, T. and Jensen, U (1999). Optimal stopping in a burn-in model. *Comm. Statist. Stochastic Models* 15, 931–951.

Karlin, S. and Rubin, H. (1956). The theory of decision procedures for distributions with monotone likelihood ratio. *Ann. Math. Statist.* 27, 272–299.

Lehmann, E. L. (1986). *Testing statistical hypotheses.* Second edition. John Wiley & Sons, New York.

Lindley, D. V. (1985). *Making decisions.* Second edition. John Wiley & Sons, Ltd., London.

Macci, D. V. (1999). Some further results about undominated Bayesian experiments. *Statist. Decisions,* 17, no. 2, 141–156.

Mosler, K. and Scarsini, M. (Eds) (1991). *Stochastic orders and decision under risk.* Institute of Mathematical Statistics, Hayward, CA.

Piccinato, L. (1980). On the orderings of decision functions. *Symposia Mathematica* XXV, Academic Press, London.

Piccinato, L. (1993). The Likelihood Principle in Reliability Analysis. In *Reliability and Decision Making* (Barlow, Clarotti, Spizzichino, Eds) Chapman & Hall (London).

Piccinato, L. (1996). *Metodi per le Decisioni Statistiche.* Springer Italia, Milano.

Runggaldier, W. J. (1993a). Concepts of optimality in stochastic control. In *Reliability and Decision Making* (Barlow, Clarotti, Spizzichino, Eds) Chapman & Hall (London).

Runggaldier, W.J. (1993b). On stochastic control concepts for sequential burn-in procedures. In *Reliability and Decision Making* (Barlow, Clarotti, Spizzichino, Eds) Chapman & Hall (London).

Savage, L. J. (1972). *The Foundations of Statistics.* Second Revised Edition. Dover, New York.

Shiryaev, A. N. (1978). *Statistical sequential analysis. Optimal stopping rules.* Springer-Verlag, New York.

Spizzichino, F. (1991). Sequential Burn-in Procedures. *J. Statist. Plann. Inference* 29 (1991), no. 1-2, 187–197.

Torgersen, E. (1994). Information orderings and stochastic orederings. In *Stochastic Orders and Their Applications* (Shaked and Shamthikumar, Eds). Academic Press, London.

van der Duyn Schouten, F. (1983). *Markov decision processes with continuous time parameter.* Mathematical Centre Tracts, 164. Mathematisch Centrum, Amsterdam.

Essential bibliography

Aldous, D.J. (1983). *Exchangeability and related topics.* Springer Series Lecture Notes in Mathematics, 1117.

Barlow, R. E. and Proschan F. (1975). *Statistical theory of reliability and life testing.* Holt, Rinehart and Winston, New York.

Bremaud, P. (1981). *Point processes and queues. Martingale dynamics.* Springer Verlag, New York.

Chow Y. S. and Teicher H. (1978). *Probability theory. Independence, interchangeability, martingales.* Springer Verlag, New York.

Cox, D. and Isham, V. (1980). *Point Processes.* Chapman & Hall, London.

Cox, D. and Oakes, D. (1984). *Analysis of Survival Data.* Chapman & Hall, London

de Finetti, B. (1970). *Teoria delle Probabilita'.* Einaudi, Torino. English translation: Theory of probability. John Wiley and Sons, New York, 1974.

De Groot, M. H. (1970). *Optimal Statistical Decisions.* McGraw-Hill, New York.

Karlin, S. (1968). *Total positivity.* Stanford University Press, Stanford, Ca.

Lawless, J. F. (1982). *Statistical models and methods for lifetime data.* John Wiley & Sons, New York.

Marshall, A.W. and Olkin I. (1979). *Inequalities: theory of majorization and its applications.* Academic Press, New York.

Savage, L. J. (1972). *The foundations of statistics,* 2nd revised edition. Dover, New York.

Shaked, M. and Shanthikumar, J.G. (1994). *Stochastic orders and their applications.* Academic Press, London.

Index

Subjective
Probability
Models for
Lifetimes

Fabio Spizzichino

Department of Mathematics
Universita' La Sapienza
Rome Italy

CRC Press
Taylor & Francis Group
Boca Raton London New York

CRC Press is an imprint of the
Taylor & Francis Group, an **informa** business

A CHAPMAN & HALL BOOK

CRC Press
Taylor & Francis Group
6000 Broken Sound Parkway NW, Suite 300
Boca Raton, FL 33487-2742

First issued in paperback 2019

ISBN-13: 978-1-58488-060-8 (hbk)
ISBN-13: 978-0-367-39717-3 (pbk)

Library of Congress Card Number 2001028129

Library of Congress Cataloging-in-Publication Data

Spizzichino, F. (Fabio), 1948-
    Subjective probability models for lifetimes / Fabio Spizzichino.
        p.   cm. -- (Monographs on statistics and applied probability ; 91)
    Includes bibliographical references and index.
    ISBN 1-58488-060-0 (alk. paper)
    1. Failure time data analysis. 2. Probabilities. I. Title. II. Series.

QA276 .S66 2001
519.5--dc21
                                                          2001028129

Visit the Taylor & Francis Web site at
http://www.taylorandfrancis.com

and the CRC Press Web site at
http://www.crcpress.com

T - #0049 - 111024 - C0 - 229/152/15 - PB - 9780367397173 - Gloss Lamination